Holiday activities

MATHS HOMEWORK

Published by Scholastic Publications Ltd,
Villiers House,
Clarendon Avenue,
Leamington Spa,
Warwickshire CV32 5PR

© 1994 Scholastic Publications Ltd
Text © 1994 University of North
London Enterprises Ltd

Activities by the IMPACT Project at the University of North London, collated and rewritten by Ruth Merttens and Ros Leather

Editors Jane Wright and Jo Saxelby-Jennings
Assistant editors Sophie Jowett and Joanne Boden
Designer Lucy Smith
Series designer Anna Oliwa
Illustrations Shaun Williams
Cover illustration Roger Wade Walker

Designed using Aldus Pagemaker
Processed by Pages Bureau, Leamington Spa
Artwork by Pages Bureau, Leamington Spa
Printed in Great Britain
by Clays Ltd, St Ives plc

British Library Cataloguing-in-Publication Data
A catalogue record for this book is
available from the British Library.

ISBN 0-590-53163-8

All rights reserved. This book is sold subject to the condition that it shall not, by way of trade or otherwise, be lent, hired out or otherwise circulated without the publisher's prior consent in any form of binding or cover other than that in which it is published and without a similar condition, including this condition, being imposed upon the subsequent purchaser.

No part of this publication may be reproduced, stored in a retrieval system, or transmitted, in any form or by any means, electronic, mechanical, photocopying, recording or otherwise, without the prior permission of the publisher. This book remains copyright, although permission is granted to copy pages 6 to 96 for classroom distribution and use only in the school which has purchased the book.

Holiday activities

Contents

Introduction	5
Half-term letter	6
Parents' booklet	7–9

Reception
One, two, three	10
Car age search	11
Shape hunt	12
Clock around!	13
Cook and play	14
Hide-away	15
Lamppost count-up!	16
Funny faces	17
Door count	18
Room jumping	19
Number writing	20
Fold up time	21

Year 1
Choose a car colour	22
Making peppermint creams	23
Making a basket for sweets	24
Make a number line	25
Number line game	26
Buy one item	27
Make a shop	28
Up to 20	29
Start with 10p	30
How much is your name worth?	31
What is 1 less than?	32
Totalling car number plates	33
Counting in twos	34

Year 2
Toy shop windows	35
Counting in fives	36
Counting in tens	37
Chocolate truffles	38
Taking a walk	39
Make a holiday diary	40
How much for your doors?	41
Purse values	42
Change from 20p	43
Check out receipts	44
Sharing between two	45
Dividing using 2p coins	46
Dividing using 5p coins	47
Dividing using 10p coins	48
Adding tens and ones	49
Ten more	50
Can you multiply?	51

Year 3
TV planner	52
Half	53
Roman numbers	54
Where are we going?	55
Shadows	56
Make some two-digit numbers	57
Sharing	58
Coin time-line	59
Using 0 to make big numbers	60
Reading car numbers	61
Arranging car numbers	62
50% off!	63

Holiday activities

Year 4
Toy quiz	64
Snakey numbers!	65
Car cricket	66
Coloured numbers	67
Guess-timate	68
Car search	69
Throwing chances!	70
Weight guesses	71
Three in a line	72
Hidden code	73

Year 5
Car football	74
Pasta count-up	75
Bluff it out!	76
Harder car cricket	77
Number search	78
Ten pence	79
Card tables	80
Scale animal	81
TV marathon	82
Unusual view	83
Cook away!	84
Racing along	85
Dicey squares	86
Word lengths	87

Year 6
Guess the number	88
Years and years	89
Metres and metres	90
Washing up!	91
Loads of money!	92–93
Dinosaur coins	94–95
Space hopping	96

impact CONTENTS

INTRODUCTION

This series of IMPACT books is designed to help you run a non-traditional homework scheme. Through the use of take-home maths activities, children can share maths with a parent/carer in the context of the home. The results of these activities then feed back into the classwork at school.

IMPACT works through the following processes:
- Teachers plan their maths for the next few weeks as usual and consider which parts might usefully be done at home.
- Teachers look through selected activities which fit in with what they are planning.
- The activities are photocopied and sent home with the children every week or fortnight.
- The results of each activity are brought back into the classroom by the children and form part of the following week's classwork.

In practice this process will be slightly different in each classroom and in each school. Teachers may adapt it to fit their own way of working and the ethos of the school in which they work.

Holiday activities

This book has been especially designed to provide teachers with activities which supply suitable mathematical tasks for parents and children to share when there is no school, such as during holiday time.

They provide:
- A selection of activities which parents will see as both mathematically and educationally valuable.
- A means of helping children to retain the skills and knowledge acquired at school over the holiday break.
- A bank of ideas to practise number facts and explore patterns in number.
- A series of graded activities in a variety of maths topics.
- A fun way to pass the time on boring car journeys and rainy days.

Selecting the activities

Unlike the usual IMPACT homework, these activities are not intended to fit back into the routine classwork a day or so later. Indeed, it may be several weeks before the children are back in class and the same teacher may not be sending the activities home and greeting the children after they have done them! Therefore, it is helpful to bear the following points in mind:
- The number of activities you send with each child will depend upon the length of the holiday. A good 'rule of thumb' is one or two per week, depending on the time of year – in poor weather, the children are more likely to be pleased to have something to do!
- It is most successful if you 'mix and match' the activities. For example, a selection involving one game, one 'making or doing' and one 'colouring' or 'completing' activity will provide a variety of opportunities to satisfy every child. A 'car' activity is a good idea if it is a time of year when people travel.
- Choose some activities which rehearse basic skills or knowledge which the children have acquired during the term and which you would like them to remember next term!
- If it is the summer holidays, you may like to consult the teacher who will be receiving the children in September, in case there is a particular topic into which one of the holiday activities could lead.
- Over half-term, send just one activity or one ordinary one and one for the car! It is better not to overload the parents who, despite the best of intentions, sometimes find that time has flown past!

Sending the activities home
Holidays

Pages 7–9 of this book can be photocopied and made up into an A5-sized booklet of advice and information for parents, to be sent home alongside a small selection of the activities.

Pages 7–9 are arranged so that you may photocopy pages 1/6 of the booklet on to the back of a cover designed by you or the children specifically for your school. The remaining booklet pages, 2/5 and 3/4, may be photocopied back to back to make the centre of the booklet.

 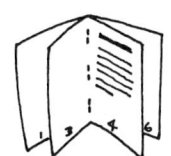

The same booklet is usually used across the whole school, so it can be made up in bulk, saving time and money.

Half-terms

Page 6 of this book is a photocopiable letter of advice for parents which can be sent home with the half-term activities.

After the holidays

Although it will not be possible to follow up these holiday activities in the same way as is normally expected with shared homework, there are a number of strategies which can be used to ensure that children and parents feel that their efforts at home have been recognised and appreciated.

- If the children have played a game, play it either 'on the rug', or in groups in class. Talk about who won and who lost. How easy, or hard, was the game? Can they make up their own version?
- If the children have made or looked for something (for example, shapes, lampposts or litres), ask how many they found and how easy it was to find them.
- Do the activities yourself at home and bring in your products to encourage the children to share theirs in the same way.
- Start the new term or half-term by writing a comment about the holiday activities into the child's shared homework or IMPACT diary so that the parents know that you know that they have done it!

Working with parents

It is important for the success of IMPACT that the activities taken home are seen by the parents to be 'proper' maths. We always suggest that activities are chosen which have an obvious mathematical purpose. Save the more 'wacky' activities for the classroom!

Each activity contains a note to parents explaining the purpose of the activity and how they can best help.

Help with implementing IMPACT

Schools that wish to get IMPACT started by means of a series of staff meetings or in-service days may like to purchase the IMPACT INSET pack which contains everything that is needed for getting going. This is available from IMPACT Supplies Ltd, PO Box 1, Woodstock, Oxon. OX20 1HB.

Useful telephone numbers

IMPACT Central Office (for information and assistance): 071 607 2789 Ext. 6349 at the University of North London.
IMPACT Supplies Ltd (for diaries and INSET pack): 0993 812895.

impact
HOLIDAY ACTIVITIES

Dear Parent or Carer

We are enclosing some maths for you to share with your child over half-term.

Please can you remember the following:

- Choose a time to do these activities when sharing the maths does not have to compete with something else particularly exciting – like a favourite film or bike race! The child will put much more energy in if they feel they are getting your attention when they are bored!

- Be as enthusiastic as you can. A gloomy attitude and a muttered, 'This looks a waste of time!', can kill a valuable activity.

- Try to let the child take the lead! Allow him or her to explain things to you if possible. It is by putting things into their own words that they learn.

- We are interested in how your child got on with the activity so do let us know by filling out the homework diary and writing a (brief) comment if you want. It is your support of your child's learning which helps them to succeed! We value all the help you give.

We hope you have fun with these activities.

Yours sincerely

....
(Class teacher)

impact
HOLIDAY ACTIVITIES

Enjoy your holiday!
Have fun with the activities! If you have any concerns, do come and tell us about these in the new term.

These activities are all designed for you to share with your child over the holiday.

Remember a few golden rules!

● **Anyone can help!**
Anyone can help your child – the family, friends (your child's and yours!), grandma, grandpa, aunties or uncles (real or 'borrowed'!), the neighbours, the cat, teddy, or even the milkman or postwoman!

● **Little and often!**
It is not good to have a day when you panic and think, 'Oh my! We haven't done any maths!' and then try to do about four hours in one day. It is good to do one activity a day, or every so often. Play the activity until the child is getting fed up and then leave it.

● **Where and when?**
Some of these holiday activities are specifically designed for car journeys. Some are best done in the kitchen, perhaps when you are cooking anyway. Some the children can do while you are peeling potatoes and some need your full and undivided attention – like a game or puzzle!

● **Make it fun!**
If you are playing a game, make it fun! Don't try too hard and don't be too clever! If the child hates losing, perhaps teddy can play as well. This lowers the temperature!

DON'T BE TOO CLEVER

How do you help?

● Choose activities suitable for the day and time.

● Spread the activities out over the holiday. Do not try to do them all in the last week!

● Don't worry if you don't get to do all the activities. We shall do lots of similar things at school!

● Listen to what your children tell you. They may be trying to make sense of something. Give them a chance!

● Talk and ask questions. Putting things into words is a vital part of learning.

● Praise again and again. Always think of something positive to say first!

The purpose of these activities is to link the maths children do in school with the maths in their home.

For example:

● **Number**
Learning to count at school is helped by practising counting at home! We can add and take away things at home, and at a later stage, we can help children to memorise their number facts through the use of games, puzzles and activities.

● **Shape**
There are shapes all around us – and used in different ways as well! Upside-down triangles, squares turned on their corners and tiling patterns all help children to recognise the different properties of the various shapes.

● **Measurement**
Children can learn to measure things at home, placing units one after another. They can also see how standard measures are used by looking at litres, grams and so on, on household goods and food packets.

● **Handling data**
If the children are to handle data in class and display and analyse it mathematically, they will need to collect their information at home! They may need to know how many cups of coffee you drink or how old the cat is!

Is this a wise move?

Yes, because:

● all the activities will extend the work done in school;

● all the activities are directly related to the National Curriculum in maths;

● the children have to use mathematical skills learned at school in different situations;

● the children may forget as much over the holiday period!

● the children may want to explain some of the maths to you. This is very good for them. It is in explaining to others what we have learned, that we really come to make that knowledge our own.

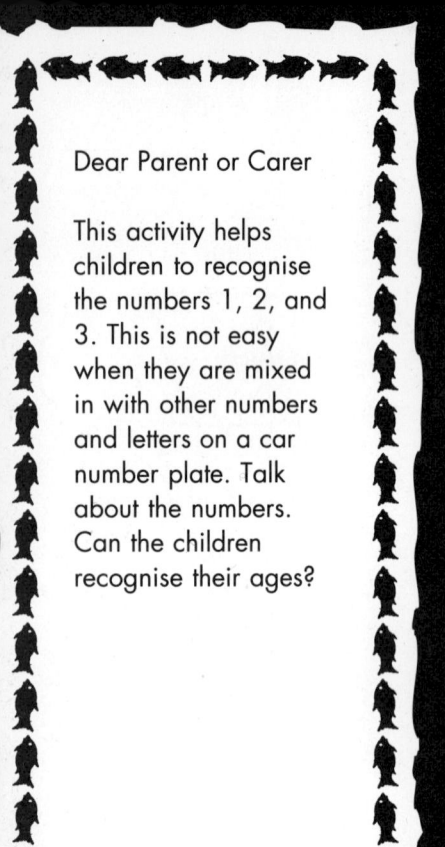

Dear Parent or Carer

This activity helps children to recognise the numbers 1, 2, and 3. This is not easy when they are mixed in with other numbers and letters on a car number plate. Talk about the numbers. Can the children recognise their ages?

_____and
child

_____
helper(s)

did this activity together

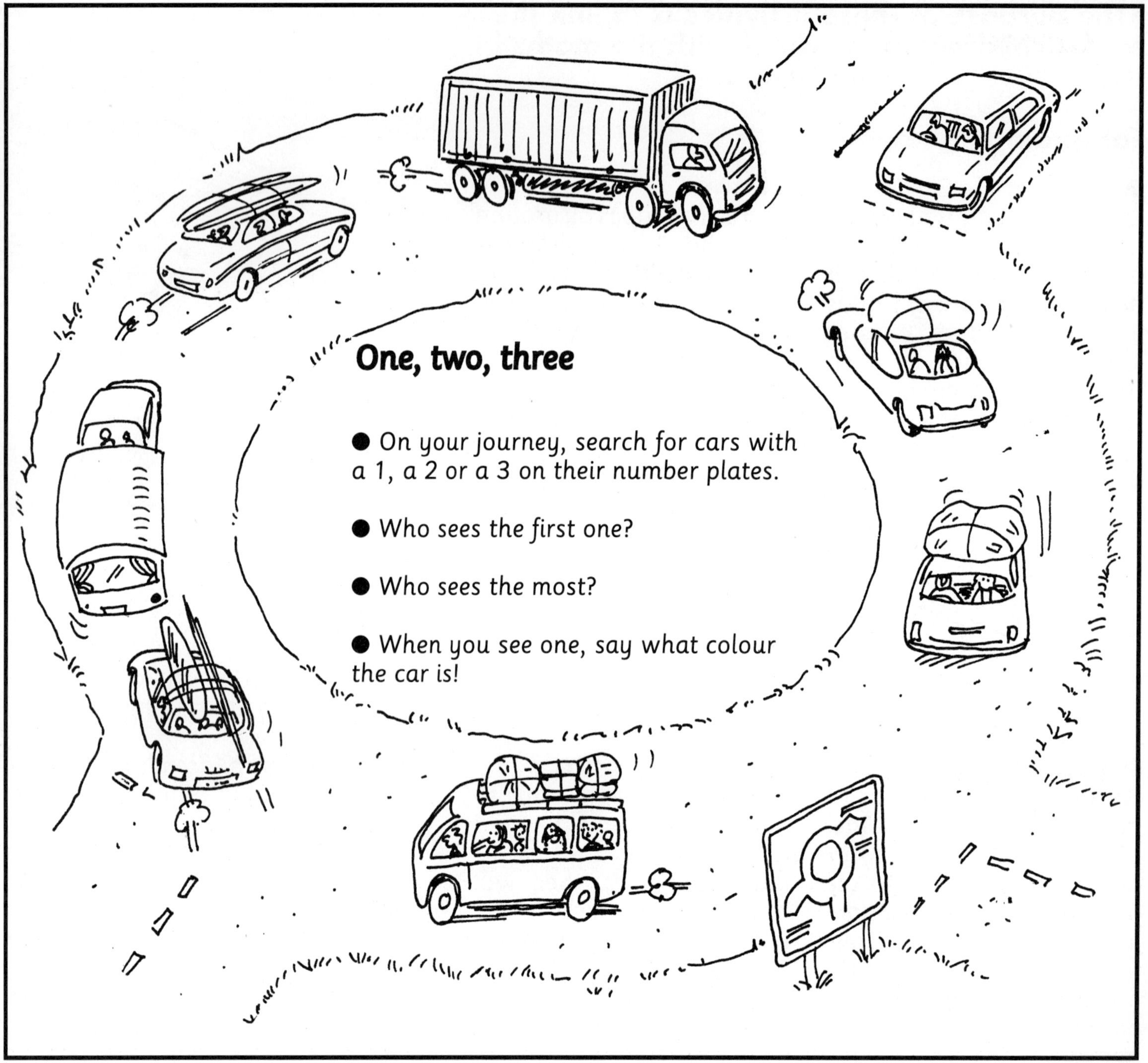

10 **Holiday activities**

One, two, three

- On your journey, search for cars with a 1, a 2 or a 3 on their number plates.

- Who sees the first one?

- Who sees the most?

- When you see one, say what colour the car is!

impact MATHS HOMEWORK

Car age search

- Hunt for your age in a car number when you are on a car journey!
- Look at as many cars as you can. Can you see your age?
- How many times do you see it?
- Can you look for someone else's age, perhaps your mum or dad?

impact MATHS HOMEWORK

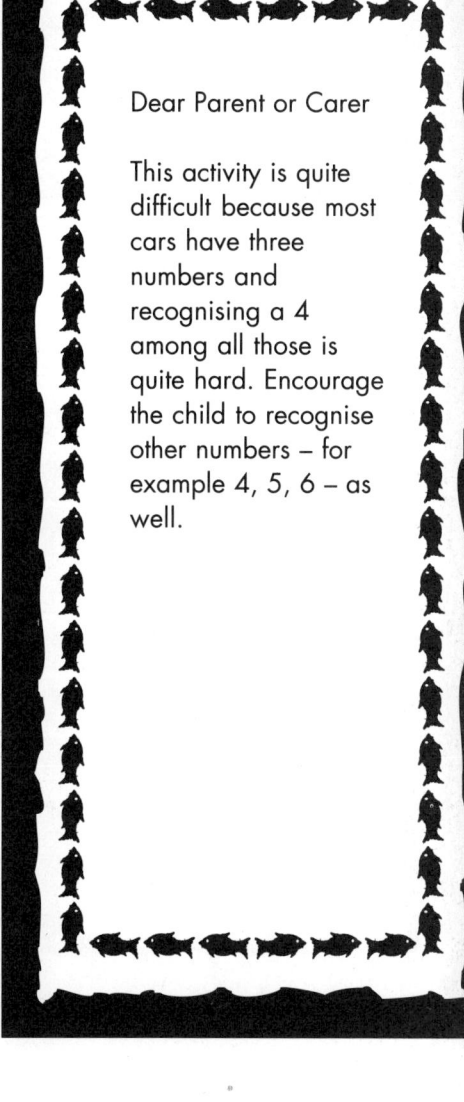

Dear Parent or Carer

This activity is quite difficult because most cars have three numbers and recognising a 4 among all those is quite hard. Encourage the child to recognise other numbers – for example 4, 5, 6 – as well.

_____ and
child

_____
helper(s)

_____

did this activity together

Holiday activities

Dear Parent or Carer

This activity helps children to look for and recognise the most common shapes. Remember that a square is still a square if it is standing on its corner! A triangle can be any way up – it only has to have three sides. An oblong is a four-sided figure with two pairs of equal sides and a right-angle in each corner. A square and an oblong are both types of rectangle!

_____ and

child

_____

helper(s)

_____

did this activity together

Shape hunt

● As you are driving along, look for an example of each of these shapes:
- a square;
- a circle;
- a triangle;
- an oblong.

● Score one point for each shape that you see. Who wins?

● Which shapes do you see most of?

12   **Holiday activities**

impact MATHS HOMEWORK

Clock around!

- Draw a big, round clock.

- Draw its numbers on its face.

- Pile that number of things (raisins, dried pasta or buttons) beside each number.

- Count around the clock. What time is it now? Point to the right hour.

- What time is teatime? Or bedtime?

- Make a pair of hands out of forks or spoons and arrange them to say 4 o'clock.

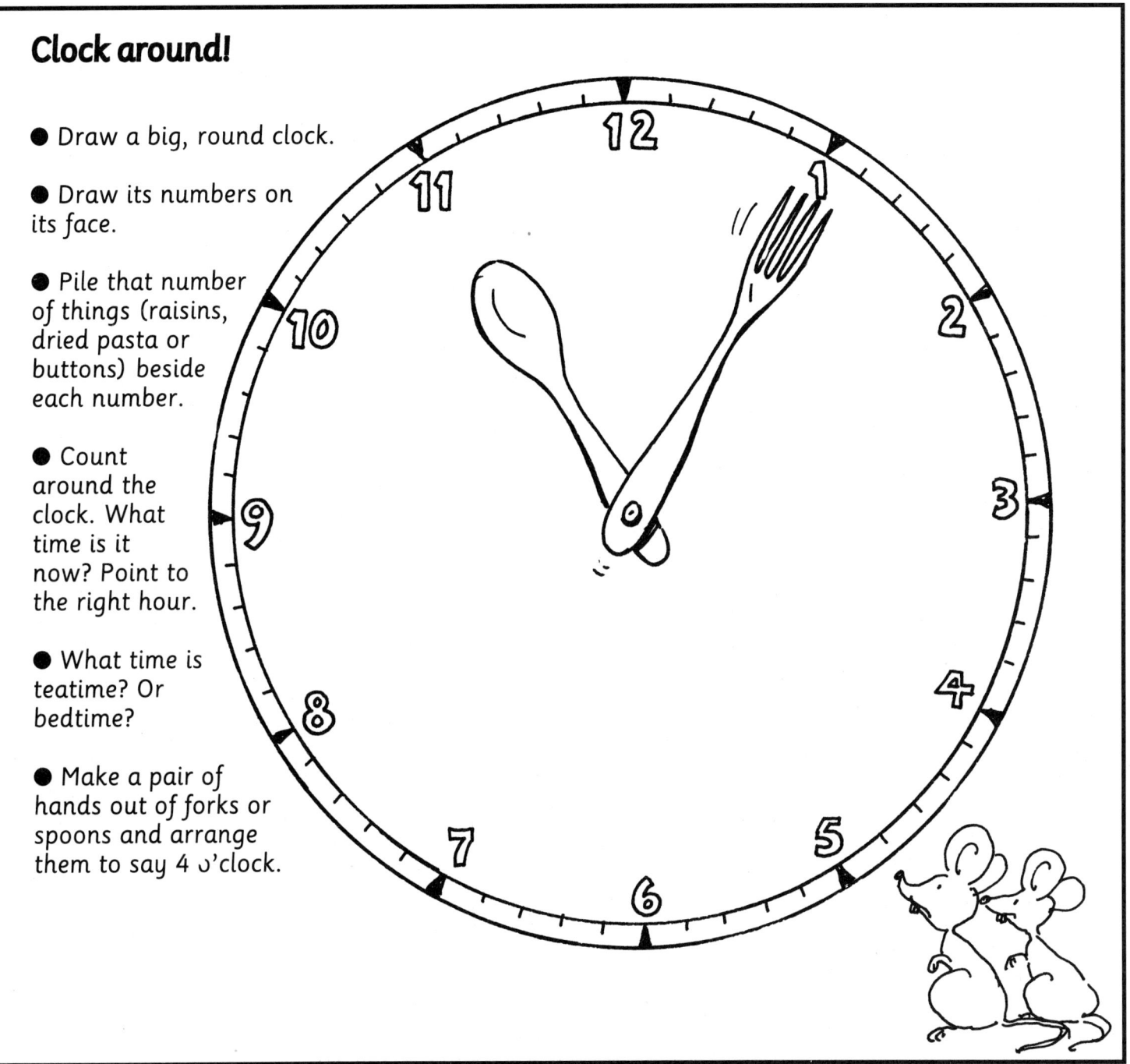

Dear Parent or Carer

Talk to the child about what happens at each different time. Stick to the hours – no half pasts or minutes at this stage – at about 8 o'clock they go to bed, or at about 5 o'clock we have our tea and so on.

_____and
child

_____
helper(s)

_____
did this activity together

Holiday activities

Dear Parent or Carer

This dough will keep for several days in an airtight polythene bag. It is entirely safe, and too salty for children to eat! Encourage the children to count the numbers of things they make and to create the numeral shapes.

_____and
child

_____
helper(s)

_____
did this activity together

14   Holiday activities

Cook and play

Our favourite dough mixture is made by mixing one cup of flour, half a cup of salt, one cup of water, one teaspoonful of cream of tartar and one dessertspoonful of oil in a pan over a low heat. You can help to stir it, until it starts to get solid. Your helper will remove it from the heat and tip it out on the table. Let it cool a bit and then knead it well.

● Make the numbers 1 to 5 from dough. Which number is your age?

● Make a pile of small balls of dough beside each number – one beside the 1, two beside the 2 and so on.

● How many balls of dough have you made in all?

impact MATHS HOMEWORK

Hide-away

YOU WILL NEED: a number of small bricks or pieces of dried pasta.

- Cut out the numbers below and make a number line.

- Put the right number of bricks or pieces of pasta beside each number.

- Take it in turns with someone to hide your eyes. While you are not looking, the other person hides one brick by moving it from one pile to another. It is then in the wrong pile!

- Can you spot which brick they have moved?

- Play it again with the other person hiding their eyes. No peeking!

1 2 3 4 5 6

Dear Parent or Carer

This game is surprisingly difficult and will involve your child in a lot of counting. Encourage them to point to each brick, or touch it, as they count it. This helps them to match the object to the word.

_____ and
child

_____
helper(s)

_____

did this activity together

Holiday activities

Dear Parent or Carer

This is a very simple form of data collection. The child is learning how to collect information on a sheet of paper and look at it later. Help them to count the number of lampposts.

_____ and
child

_____
helper(s)

_____
did this activity together

Holiday activities

Lamppost count-up!

● Next time you go shopping count how many lampposts you see.

● A really good way to do this is to take this sheet of paper and put a coloured mark on each lamppost on this page each time you see one.

● When you get home, count how many you have seen.

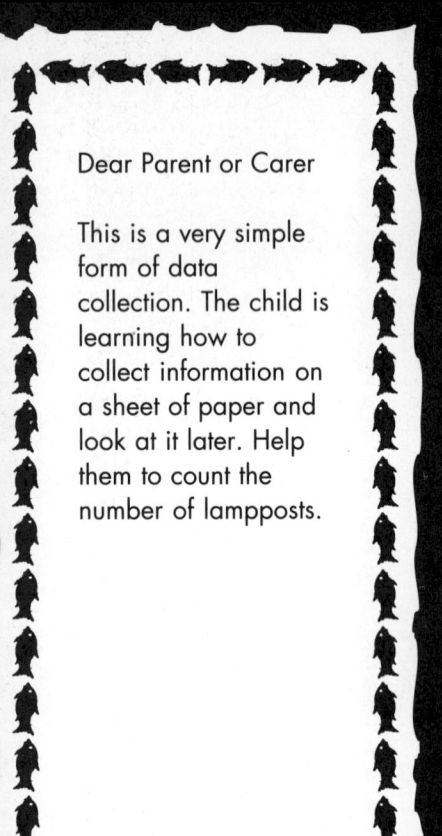

impact MATHS HOMEWORK

Funny faces

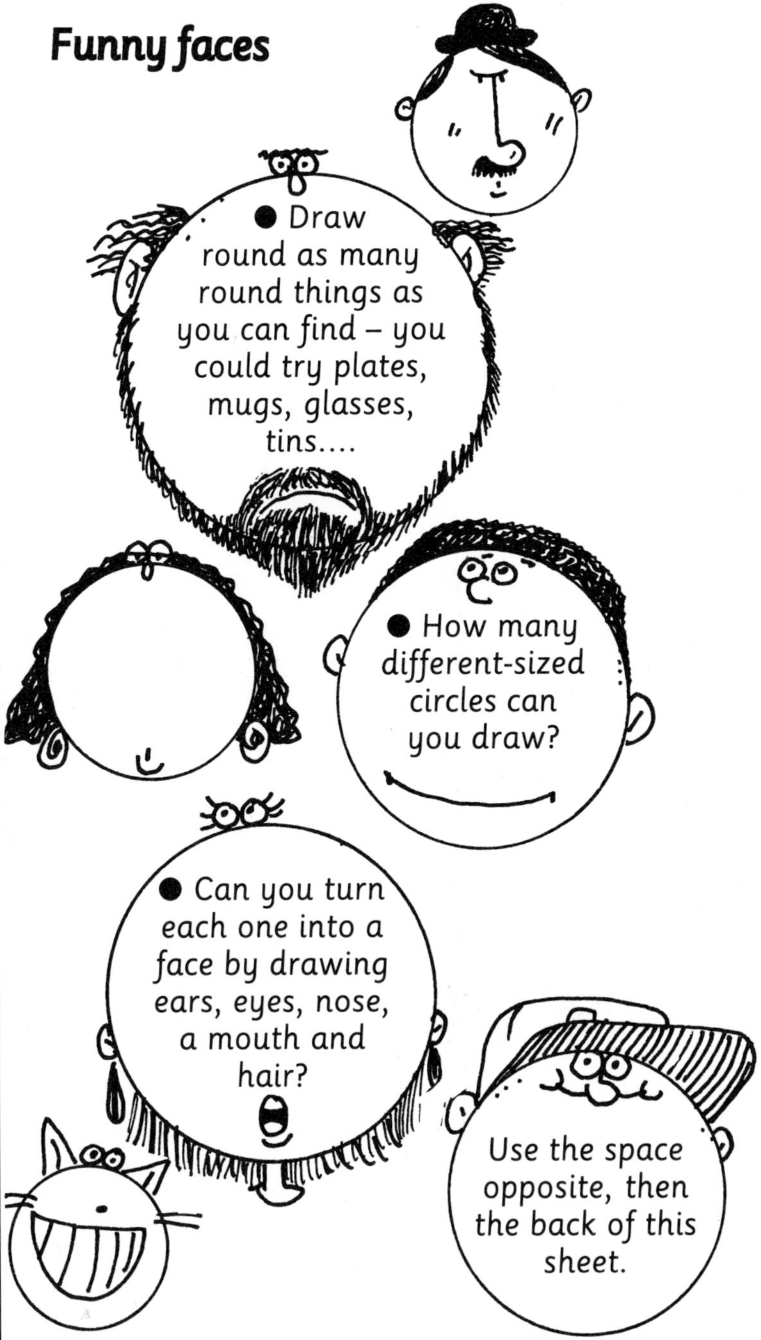

- Draw round as many round things as you can find – you could try plates, mugs, glasses, tins....

- How many different-sized circles can you draw?

- Can you turn each one into a face by drawing ears, eyes, nose, a mouth and hair?

Use the space opposite, then the back of this sheet.

impact MATHS HOMEWORK

Dear Parent or Carer

This activity helps the child's hand-to-eye co-ordination skills in drawing round something. It also reinforces the idea of a circle! Help the child by holding each object still while they draw around it.

_____ and
child

_____
helper(s)

_____
did this activity together

Holiday activities

Dear Parent or Carer

If there are not many full-sized doors, count cupboard doors as well. Count as many doors as you think your child can manage. Can they count them from memory while sitting at the kitchen table?

_____ and
child

_____
helper(s)

_____
did this activity together

18   **Holiday activities**

Door count

- How many doors can you count in your home? Count as many as you can.

impact MATHS HOMEWORK

Room jumping

- How many jumps does it take you to cross the room?

- Ask someone to count while you jump! No cheating by walking or skipping between jumps!

- Write down the number.

- Now ask someone else in your home to jump across the room.

- How many jumps do they take? You count while they jump.

- Write down the number.

- Who took the most jumps?

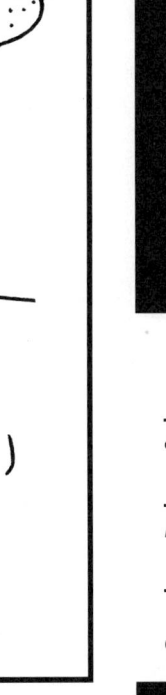

Dear Parent or Carer

This activity helps your child to count accurately and then write down the number. They may have difficulty with writing the number. A good trick is to write it very faintly and let them go over it with a felt-tipped pen.

_____and
child

_____
helper(s)

_____
did this activity together

Holiday activities

Dear Parent or Carer

It is difficult for children to learn to write numbers – help them by writing them very faintly and letting them write over your number in felt-tipped pen. Encourage them to start writing the numbers at the top.

_____ and
child

_____
helper(s)

_____
did this activity together

20 Holiday activities

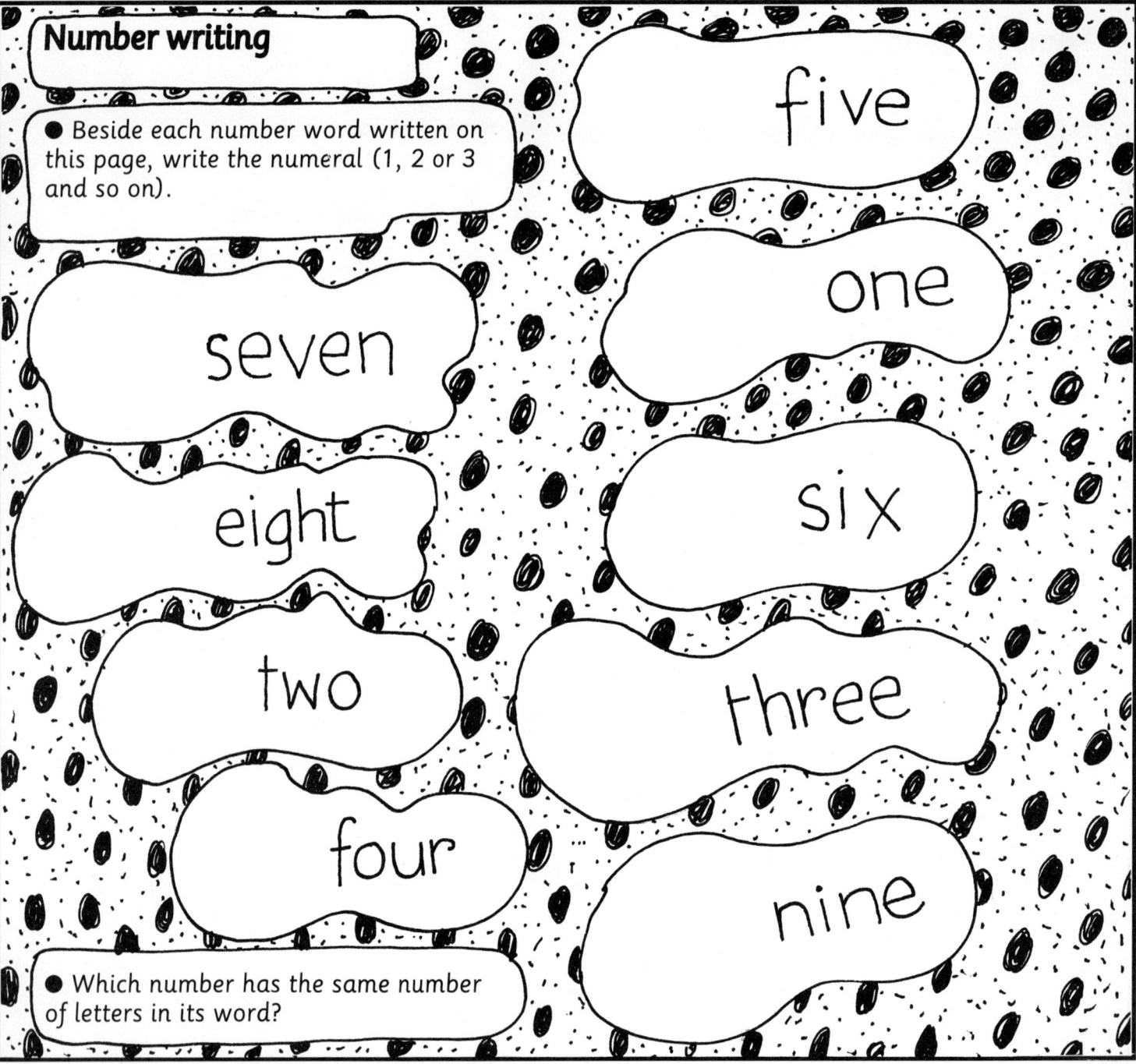

Number writing

• Beside each number word written on this page, write the numeral (1, 2 or 3 and so on).

five
one
seven
six
eight
two
three
four
nine

• Which number has the same number of letters in its word?

impact MATHS HOMEWORK

Fold up time

- Take a piece of paper and fold it in half.

- Fold it in half again.

- Cut a pie-shaped wedge in the corner.

- When you unfold it, how many holes do you think there will be in the paper? Guess first! Unfold it. Were you right?

- Try folding another piece of paper in half three times! How many holes this time?

Dear Parent or Carer

This activity makes children think about spatial arrangements and numbers. This is quite difficult for a small child who will often forget the first fold! You may need to do it several times.

_____and
child

_____
helper(s)

_____
did this activity together

Dear Parent or Carer

If a certain colour always wins, perhaps that colour can be eliminated or the game reversed so that the loser is the one who reaches 20 first. This game will encourage your child to be thinking about probability.

_____and
child

_____
helper(s)

_____
did this activity together

Holiday activities

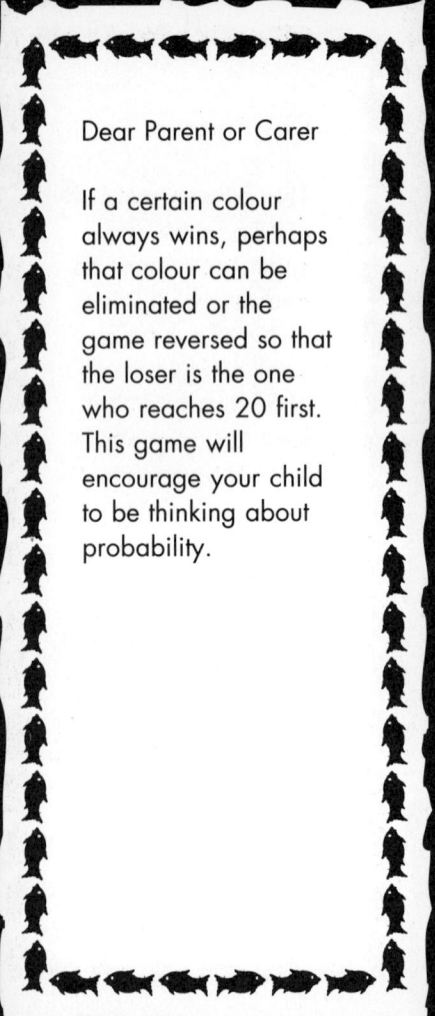

Choose a car colour

Everyone in the car has to choose a different colour.

● Count the cars which are of your chosen colour. The first person to reach 20 cars is the winner.

● Which is the best colour to choose?

● Which colours always lose?

Making peppermint creams

YOU WILL NEED: an egg, 225g of icing sugar, 2 drops of peppermint essence, 2 or 3 drops of cake colouring (optional – green is traditional), and someone to help you with this recipe.

METHOD

- Separate the egg.

- Sift the icing sugar into a bowl.

- Add a little of the egg white and carefully mix it into the icing sugar. Keep adding the egg white; be careful, as the mixture should remain stiff. Use your hands to knead the mixture, if you prefer.

- When the mixture is well blended, add the peppermint essence. Blend it in carefully.

- If you are adding colouring, divide your mixture so that some of the sweets remain white.

- Carefully shape your sweets into small round shapes.

- Decorate the top with a fork mark.

- Place the sweets on to some greaseproof paper or foil. They will be ready to eat tomorrow.

Dear Parent or Carer

Making simple sweets is a super activity. These peppermints make lovely presents – it is always a pleasure to receive a home-made gift. For the more adventurous, dip your peppermint creams into some melted chocolate.

Use your egg yolk in an omelette or for some scrambled egg.

_____and

child

_____

helper(s)

_____

did this activity together

Dear Parent or Carer

Please help your child with measuring, and perhaps drawing a line, where the fold should be. Children love making these little baskets. They can be decorated in many ways – allow your child to experiment.

They make special personal presents for Mother's Day or grandparents' birthdays.

_____ and
child

_____
helper(s)

_____
did this activity together

Making a basket for sweets

YOU WILL NEED: some card – an old cereal box will do – and some wrapping paper (old birthday or Christmas paper is very good).

● From the card, cut out a square shape 10cm by 10cm and a strip 10cm by 2cm.

● Cover each side of your square and strip with pretty paper.

● Fold each side of your square about 2cm into the centre.

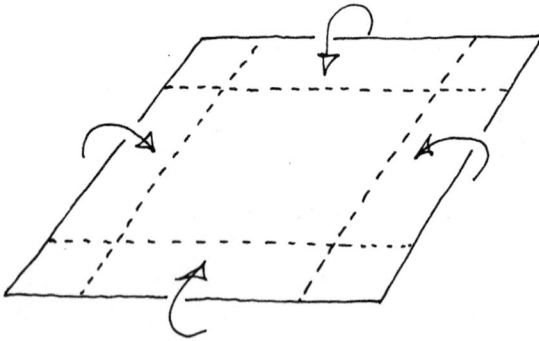

● Cut into each corner like this:

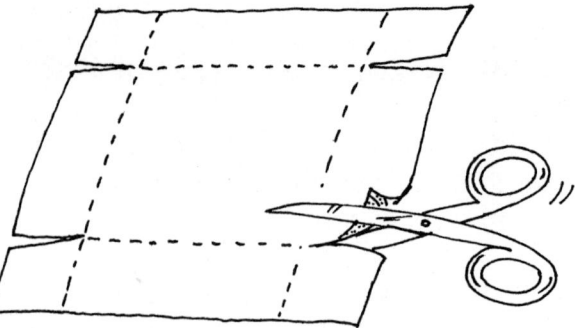

● Fold the corners round to make a small box – stick or staple it together.

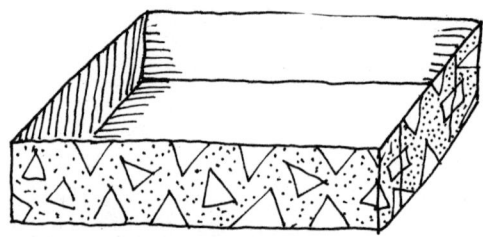

● Use the strip to make a handle.

● Fill your basket with little sweets or put in a small gift to hang on to the Christmas tree or to give as a present.

Make a number line

YOU WILL NEED: old birthday or Christmas cards, a piece of string, clothes pegs and two chairs.

- On the cards, write the numbers from 0–20.
- Make a line by tying a piece of string between the chairs.
- Fasten the numbers to the line with clothes pegs.
- How quickly are you able to arrange your numbers on the line?

Dear Parent or Carer

Many fun games can be played with number lines. Take turns to remove cards while the other person is not looking. How quickly can the missing numbers be found? Encourage your child to count up and down the number line. You may like to begin this exercise by starting in different places.

_____ and

child

_____

helper(s)

_____

did this activity together

Holiday activities

Dear Parent or Carer

This is an excellent activity for helping children to become familiar with the number line and the position of numbers. You can play this game by using two cards to start, for example 7 and 14 – this is a lot more difficult!

_____and
child

_____
helper(s)

_____
did this activity together

Number line game

YOU WILL NEED: cards numbered 0–20 made from old birthday cards.

● Place all the cards, except 10, face down in a pile.

● Turn each card over in turn.

● If the card turned over is next to 10 (that is, 11 or 9), the child must shout, 'Stop!' That card then gets placed in the number line.

● As each card is turned over, if it can go next to a number already in the line the child must shout, 'Stop!' Keep going until the number line is complete.

26   **Holiday activities**

Buy one item

● Choose which coins you would need to buy one item in the shopping trolley.

Dear Parent or Carer

Children love being responsible for paying when shopping. This is a task that they can help with on each shopping expedition. Give your child time to find the correct coins before it is your turn at the checkout.

_____ and
child

_____
helper(s)

_____

did this activity together

Holiday activities

Dear Parent or Carer

Children love playing shops. Encourage your child to have the correct money to come to the shop – giving change is difficult for young children. Ensure that the prices reflect your child's maturity.

_____and
child

_____
helper(s)

_____
did this activity together

Make a shop

- Arrange some of your toys and label them with prices.

- Use real money to play shops with a friend.

28   **Holiday activities**   *impact* MATHS HOMEWORK

Up to 20

- On a journey, look at car number plates. Concentrate on the last two digits.

- Find numbers from 0–20 in order. This is a fun game for all the family!

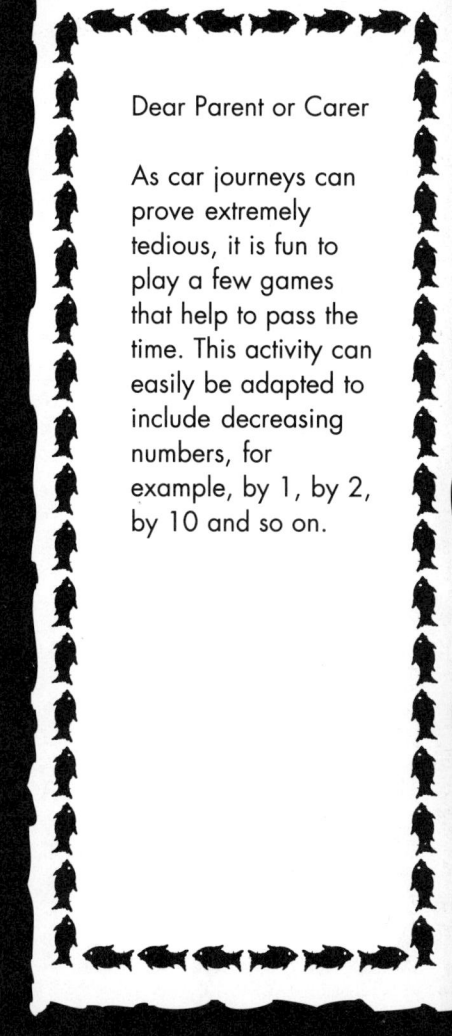

Dear Parent or Carer

As car journeys can prove extremely tedious, it is fun to play a few games that help to pass the time. This activity can easily be adapted to include decreasing numbers, for example, by 1, by 2, by 10 and so on.

_____and
child

_____
helper(s)

_____

did this activity together

impact MATHS HOMEWORK

Holiday activities

Dear Parent or Carer

Please encourage your child to begin counting on from 10; that is, 10, 11, 12 and so on. Ten is very significant in our number system. Money helps children to differentiate between tens and units (ones).

_____ and
child

_____
helper(s)

_____
did this activity together

Holiday activities

Start with 10p

- Price some of your toys with teen numbers, like the ones below.

- Always start with 10p and count upwards to find the value of the toys in your shop.

- Try these first.

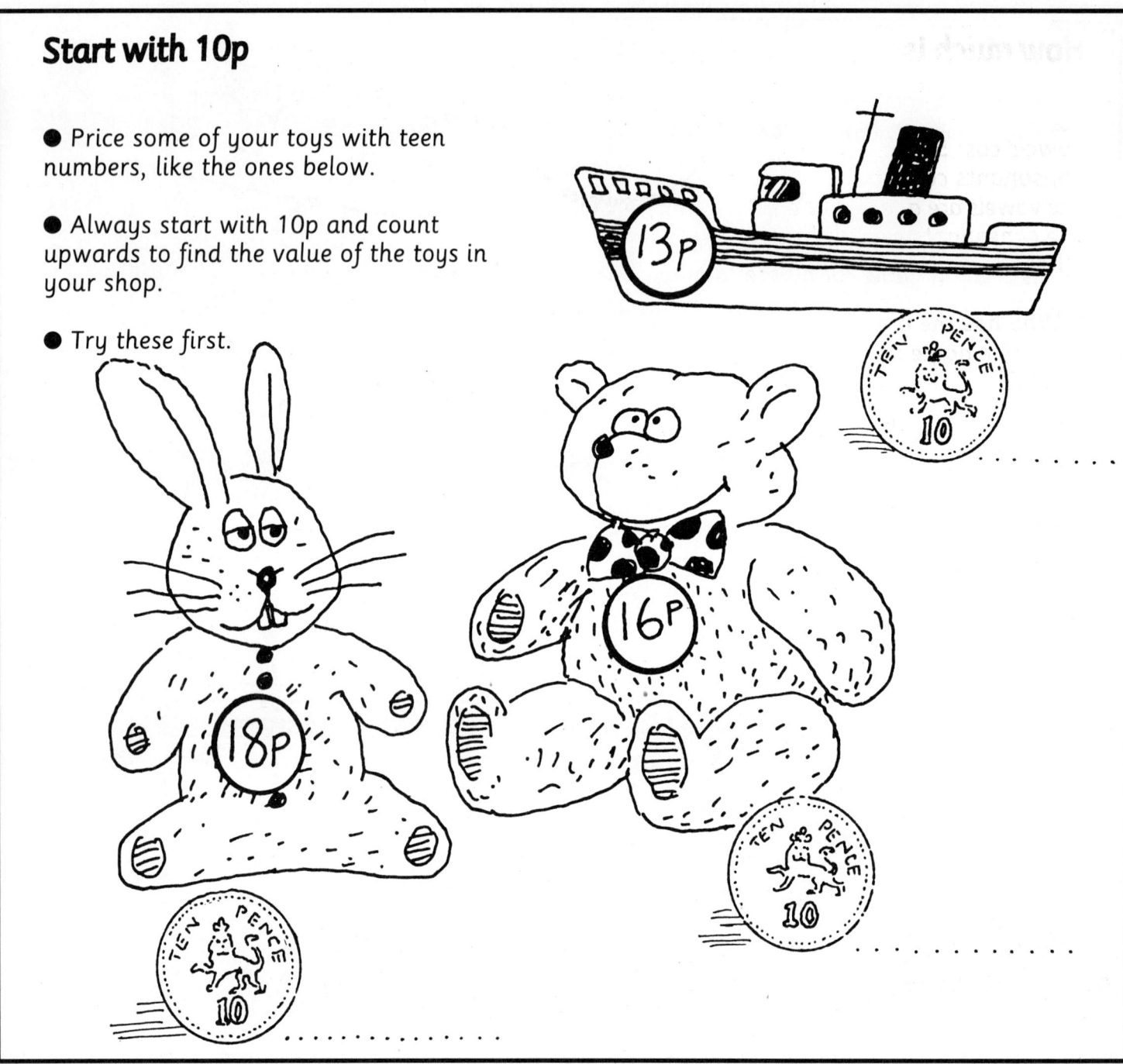

impact MATHS HOMEWORK

How much is your name worth?

Vowels cost 5p.
Consonants cost 2p.
The vowels are a, e, i, o and u.
The consonants are all the other letters in the alphabet.

- Who has the most expensive name in your family?

- Can you arrange the names in value order?

- How much do all the names in your family cost?

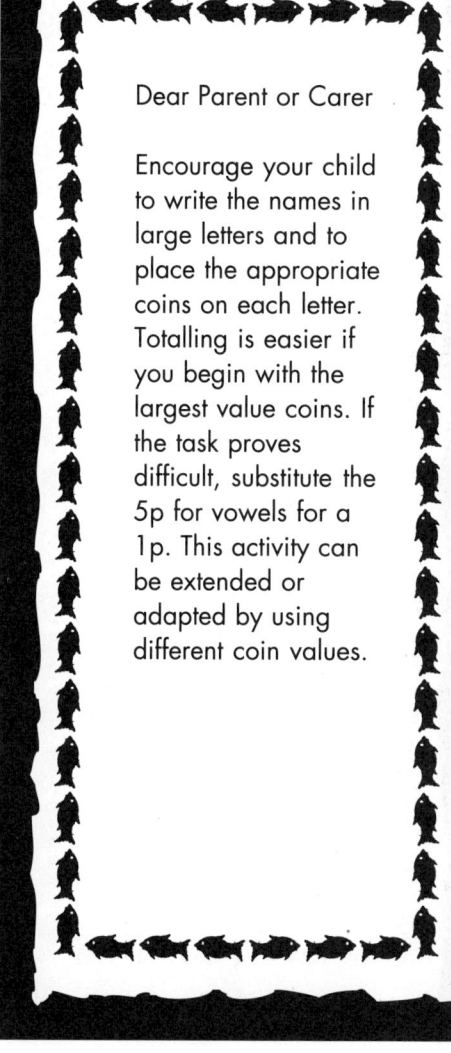

Dear Parent or Carer

Encourage your child to write the names in large letters and to place the appropriate coins on each letter. Totalling is easier if you begin with the largest value coins. If the task proves difficult, substitute the 5p for vowels for a 1p. This activity can be extended or adapted by using different coin values.

_____and
child

_____
helper(s)

_____
did this activity together

Holiday activities

Dear Parent or Carer

Children often find it difficult to be able to reduce a number by 1. This is especially true for the tens numbers – 10, 20, 30, 40 and so on. This activity can be adapted to 2 less or even 3 less.

_____and
child

_____
helper(s)

_____
did this activity together

32 Holiday activities

What is 1 less than?

This is a game to play on a journey.

One person chooses a number, then the others take turns to reduce the number by 1; for example, 40, 39, 38 and so on.

impact MATHS HOMEWORK

Totalling car number plates

- How quickly can you add up the digits on car number plates? Try the one below first, before trying this out on a car journey.

- On your journey, take turns to find the totals. Who got the biggest total? Who got the smallest total?

- Try again.

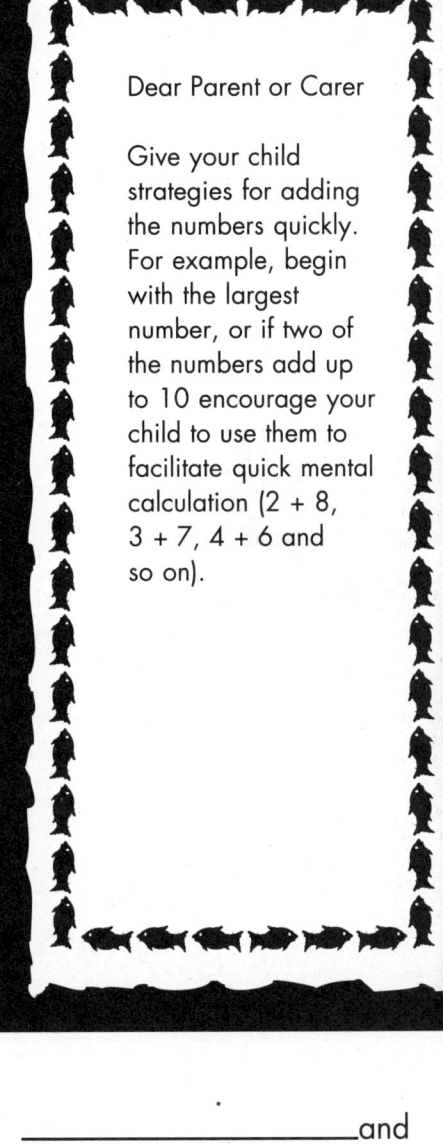

Dear Parent or Carer

Give your child strategies for adding the numbers quickly. For example, begin with the largest number, or if two of the numbers add up to 10 encourage your child to use them to facilitate quick mental calculation (2 + 8, 3 + 7, 4 + 6 and so on).

_____and
child

_____
helper(s)

_____
did this activity together

Dear Parent or Carer

This will help your child to learn the two times table. Please encourage them to estimate the answer before checking. It may be easier to begin this activity with fewer 2p coins as this may increase confidence.

_____and
child

_____
helper(s)

_____

did this activity together

34   Holiday activities

Counting in twos

YOU WILL NEED: 10 × 2p coins laid out as below.

- Can you count the coins in twos?

5 lots of 2p are ....

8 lots of 2p are ....

Five lots of 2p could look like this.

impact MATHS HOMEWORK

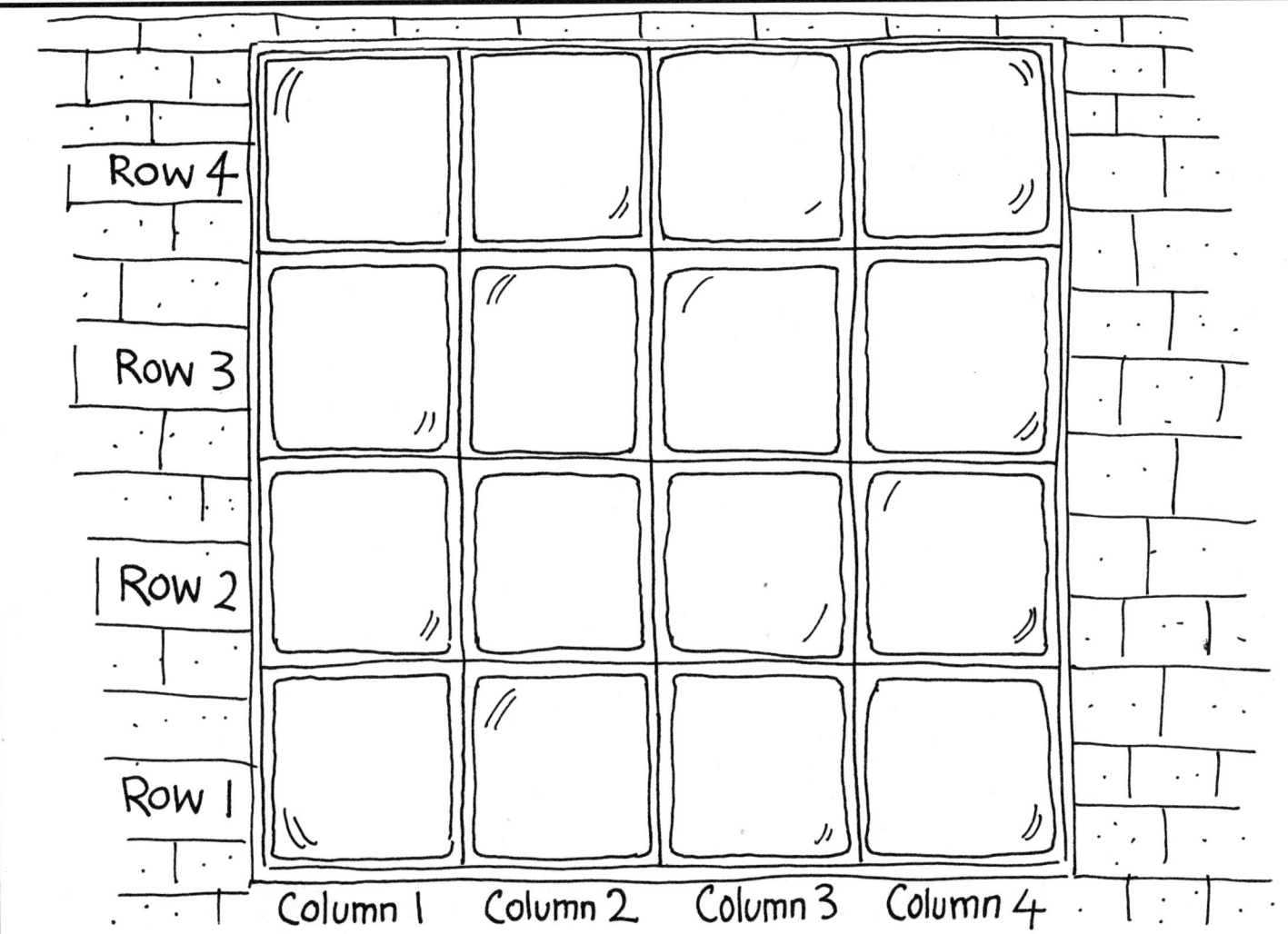

Toy shop windows

● Draw small pictures of some of your favourite toys (or cut out pictures from a magazine) and place them in the squares of the toy shop window above.

● Play a game like this with a friend: 'My favourite cuddly toy is in Column 3, Row 2. What is it?'

● Alternatively, take the toys out of the window and ask someone to put, say, a teddy in Column 3, Row 4.

Dear Parent or Carer

This activity will help your child with coordinates, grid work and map reading. Help your child to remember that it is always columns first and then rows ('Along the corridor and up the stairs' is a useful way to remember this rule).

_____and
child

_____
helper(s)

_____

did this activity together

Holiday activities

Dear Parent or Carer

This activity will help your child to learn the five times table. Encourage your child to estimate the answer before checking. You may begin with fewer 5p coins to give your child confidence.

Many children like to imagine that each digit of their hand is worth 5p. If you show three digits they will represent 5p, 10p, 15p, and so on.

_____ and
child

_____
helper(s)

_____
did this activity together

Counting in fives

YOU WILL NEED: 10 × 5p coins laid out as below.

● Can you count the coins in fives?

5 lots of 5p are ....

8 lots of 5p are ....

An easy way of remembering your five times table is to know that two 5s are 10, four 5s are 20 and so on.

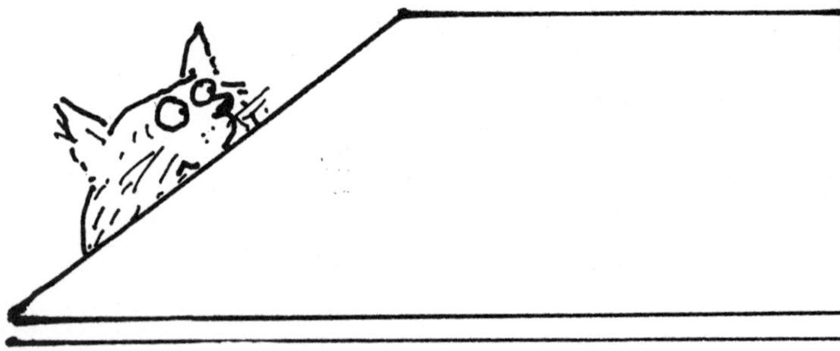

36 Holiday activities

Counting in tens

YOU WILL NEED: 10 × 10p coins laid out as below.

- Can you count in tens?

5 lots of 10p are ....

8 lots of 10p are ....

Dear Parent or Carer

This activity will help your child to learn the ten times table. Encourage your child to estimate the answer before checking. It may be easier to begin this activity with fewer 10p coins to increase confidence.

_____ and
child

_____
helper(s)

did this activity together

impact MATHS HOMEWORK

Holiday activities

Dear Parent or Carer

These sweets make a lovely present. The egg whites can be used for meringues or just added to an omelette or to scrambled egg mixture.

_____ and
child

_____
helper(s)

_____
did this activity together

38 **Holiday activities**

Chocolate truffles

YOU WILL NEED: 120g of plain chocolate (cooking or ordinary), 2 egg yolks, 3 heaped tablespoons of icing sugar, 60g of margarine, chocolate strands (optional) and someone to help you.

- With your helper, gently melt the chocolate in a basin over a pan of boiling water.

- Cream together the margarine, icing sugar and egg yolks.

- Carefully add the melted chocolate and stir the mixture. If the mixture is sticky, place in the fridge overnight.

- Roll the truffle mixture into little spheres and dip them into chocolate strands. Place them on to some foil circles and leave in the fridge to harden. Enjoy your sweets!

impact MATHS HOMEWORK

Taking a walk

- Talk to someone at home about a short walk you can take together.

- Now draw a map to show the route of your walk. Try to remember the main things that you saw, so that you can put them on your map.

Dear Parent or Carer

Encourage your child to observe carefully the static objects you see on your walk, for example the big tree on the corner or the post-box halfway along Cherry Lane. Look at the street names and try to remember them in the right order. Draw the map together. You may like to write the names of the streets on small pieces of paper that can be positioned along your map in the right order.

_____and

child

_____

helper(s)

_____

did this activity together

Holiday activities

Dear Parent or Carer

Children enjoy recording their achievements or recording something about their day out. Allow your child to stick in postcards or brochures and wrappers, to draw maps and draw round coins to indicate the cost of things.

_____ and
child

_____
helper(s)

did this activity together

Make a holiday diary

● Keep a mathematics scrapbook for a whole week. Every day write down/draw/stick something in your diary.

For example, 'On Monday I drew the coins used to visit the cinema.' Stick into the diary your cinema ticket and a picture of the film you saw.

On Tuesday we travelled from .... to .... it was .... miles
On Wednesday it took us .... minutes to eat our dinner. It was ....
On Thursday I threw the ball to .... and caught it .... times without dropping it.
On Friday I bought some chocolates, they cost £ ....

Monday
Tuesday
Wednesday
Thursday
Friday
Saturday
Sunday

Holiday activities

impact MATHS HOMEWORK

How much for your doors?

Upstairs doors cost 6p.
Downstairs doors cost 4p.

- Draw all the doors in your home and place the coins of that value on your pictures.

- How much do all the upstairs doors put together cost?

- Do all the downstairs doors cost more or less than all the upstairs doors?

- How much do all the doors in your house cost?

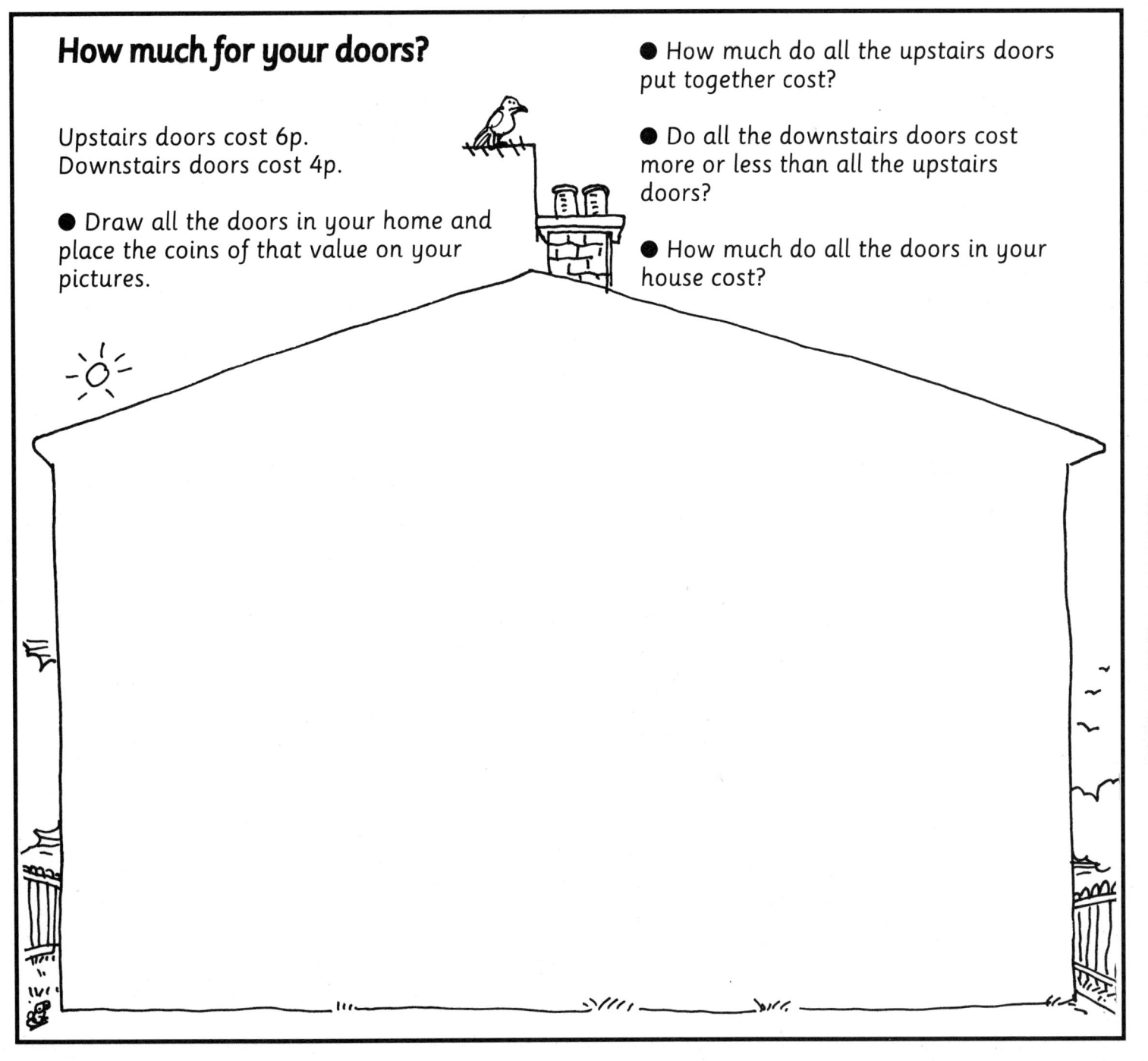

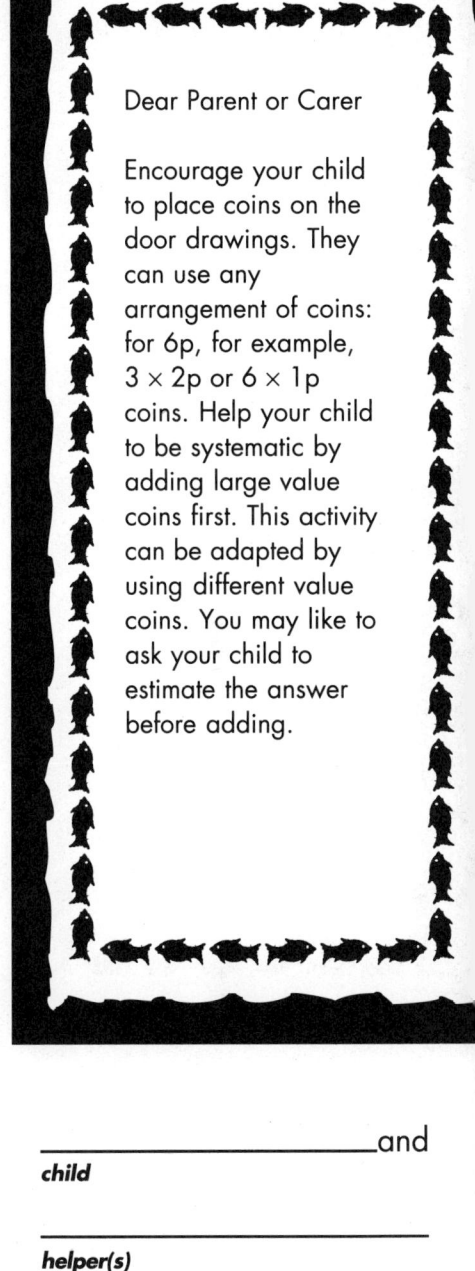

Dear Parent or Carer

Encourage your child to place coins on the door drawings. They can use any arrangement of coins: for 6p, for example, $3 \times 2p$ or $6 \times 1p$ coins. Help your child to be systematic by adding large value coins first. This activity can be adapted by using different value coins. You may like to ask your child to estimate the answer before adding.

_____ and
child

_____
helper(s)

_____
did this activity together

impact MATHS HOMEWORK

Holiday activities

Dear Parent or Carer

Encourage your child to arrange the coins in value order before trying to calculate the answer. This is an excellent activity for facilitating quick mental arithmetic. Encourage your child to estimate the amount before beginning. You could adapt this activity so that your child experiences a variety of different calculations.

_____ and
child

_____
helper(s)

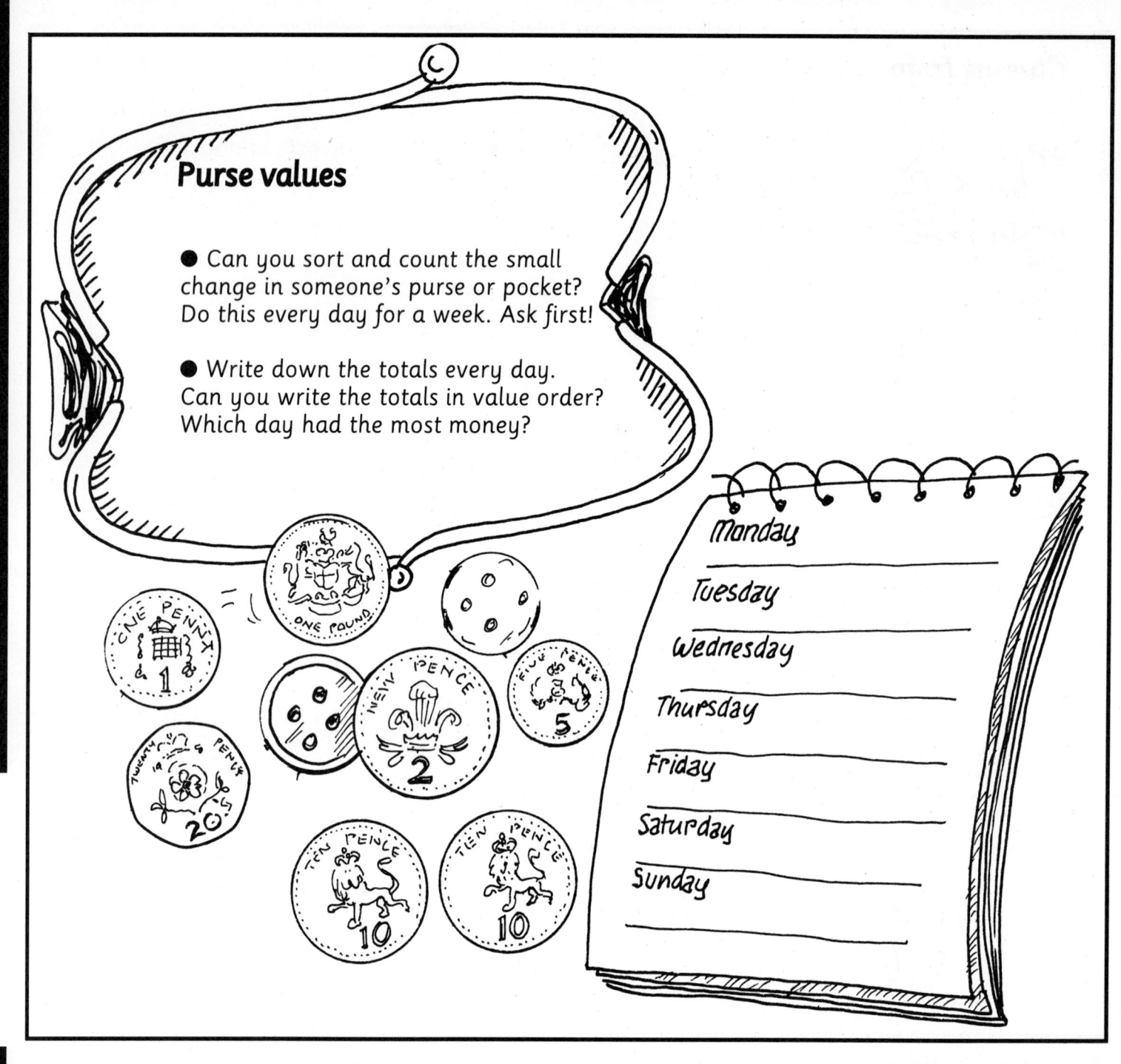

_____
did this activity together

Holiday activities

Purse values

- Can you sort and count the small change in someone's purse or pocket? Do this every day for a week. Ask first!

- Write down the totals every day. Can you write the totals in value order? Which day had the most money?

Monday
Tuesday
Wednesday
Thursday
Friday
Saturday
Sunday

impact MATHS HOMEWORK

Change from 20p

Prices on toys: 5p (jigsaw), 15p (teddy), 8p (car), 2p, 12p (football), 16p (baby), 19p (truck).

- Label some of your toys with prices less than 20p. Play shops with a friend. (You will each need 20p to take to the shop.)

- Take turns to buy an item, making sure you get the correct change.

Dear Parent or Carer

Please help your child with counting on when giving change; for example, 15p for the teddy: '16, 17, 18, 19, 20 – the change is 5p.' Begin with 10p if your child finds this difficult. Many children become confused when they receive more coins, in change, after having spent money. Extend this activity to 50p. If you give 50p for a 25p item, change is often given by counting, '26, 27, 28, 29, 30, 40, 50 – the change is 25p.'

_____and
child

_____
helper(s)

_____

did this activity together

impact MATHS HOMEWORK

Holiday activities 43

Dear Parent or Carer

Please use real coins to place by the chosen prices. Give your child time to count out the required coins. It is often helpful to begin with large value coins as this makes it easier for your child to calculate the cost.

_____and
child

_____
helper(s)

_____
did this activity together

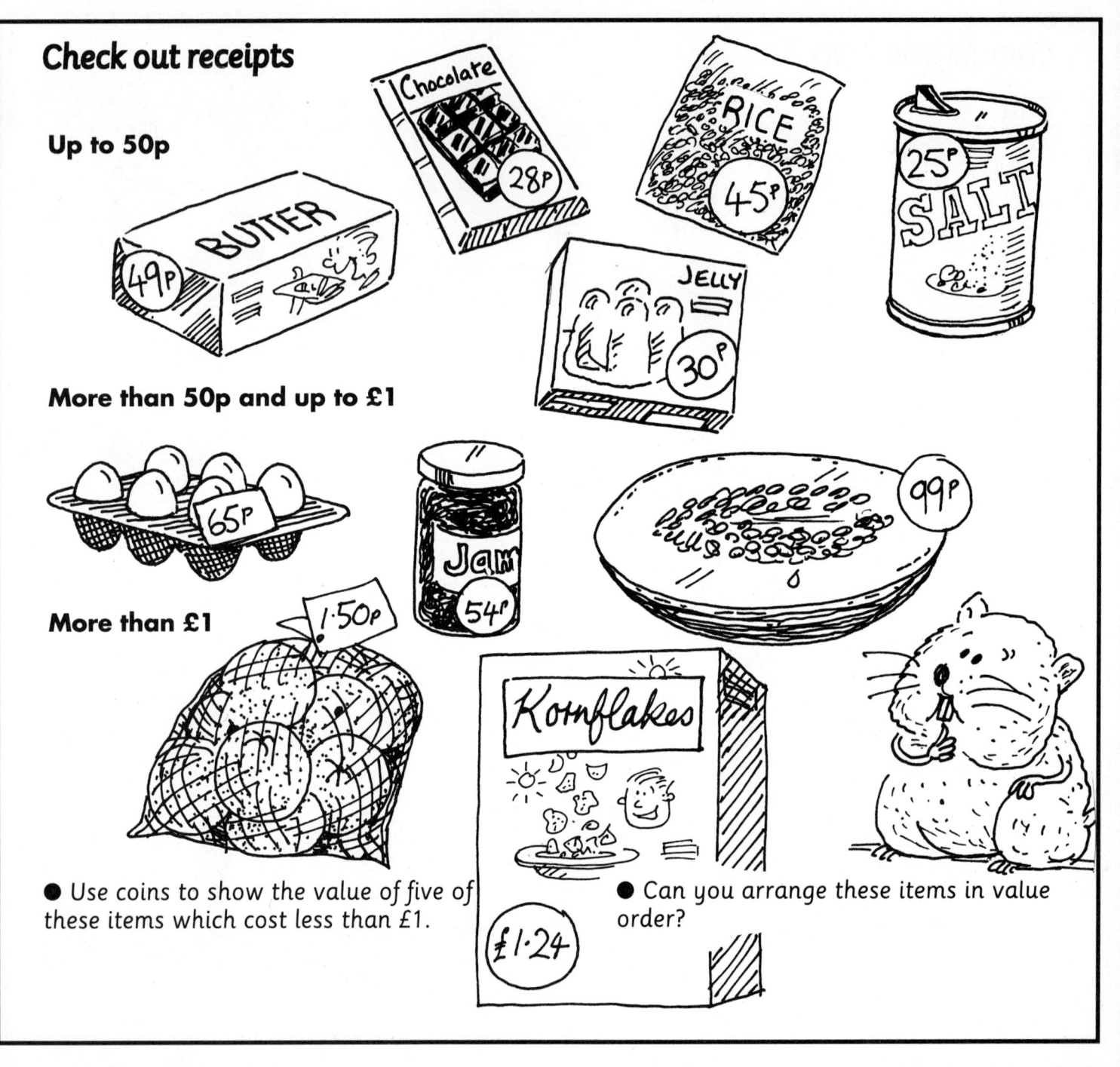

44 Holiday activities

Check out receipts

Up to 50p

- Butter 49p
- Chocolate 28p
- Rice 45p
- Salt 25p
- Jelly 30p

More than 50p and up to £1

- Eggs 65p
- Jam 54p
- (bowl) 99p

More than £1

- (potatoes) 1·50p
- Kornflakes £1·24

● Use coins to show the value of five of these items which cost less than £1.

● Can you arrange these items in value order?

impact MATHS HOMEWORK

Sharing between two

YOU WILL NEED: some raisins, pasta, counters or LEGO bricks and someone to play with.

- Begin with two raisins – if you share them between the two of you, how many do you get each?

- If you have three raisins, how can you share these between two people?

- Continue with four, five and six raisins – can you see a pattern?

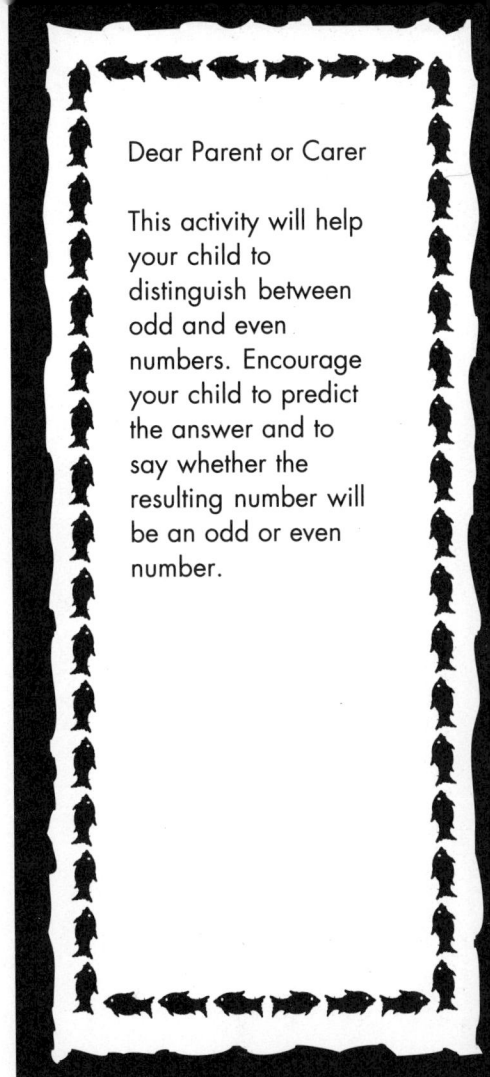

impact MATHS HOMEWORK

Dear Parent or Carer

This activity will help your child to distinguish between odd and even numbers. Encourage your child to predict the answer and to say whether the resulting number will be an odd or even number.

_____and
child

_____
helper(s)

_____

did this activity together

Holiday activities

Dear Parent or Carer

This is an excellent way of learning how to divide into twos. Encourage your child to estimate the answer, then to count out the coins to check the accuracy of their estimate.

_____ and
child

_____
helper(s)

_____
did this activity together

46   Holiday activities

Dividing using 2p coins

YOU WILL NEED: 20p in 2p coins.

● How many 2p coins make 10p? Write your answer on the coin below.

● How many 2p coins make 12p? Write your answer on the coin below.

impact MATHS HOMEWORK

Dividing using 5p coins

YOU WILL NEED: 50p in 5p coins.

- How many 5p coins make 25p?
- How many 5p coins make 15p?
- How many 5p coins make 35p?
- How many 5p coins make 50p?

impact MATHS HOMEWORK

Dear Parent or Carer

This is an excellent way of learning how to divide into five. Encourage your child to estimate the answer, then to count out the coins to check the accuracy of their estimate.

_____and
child

_____
helper(s)

_____

did this activity together

Holiday activities 47

Dear Parent or Carer

This is an excellent way of learning how to divide into ten. Encourage your child to estimate the answer, then to count out the coins to check the accuracy of their estimate.

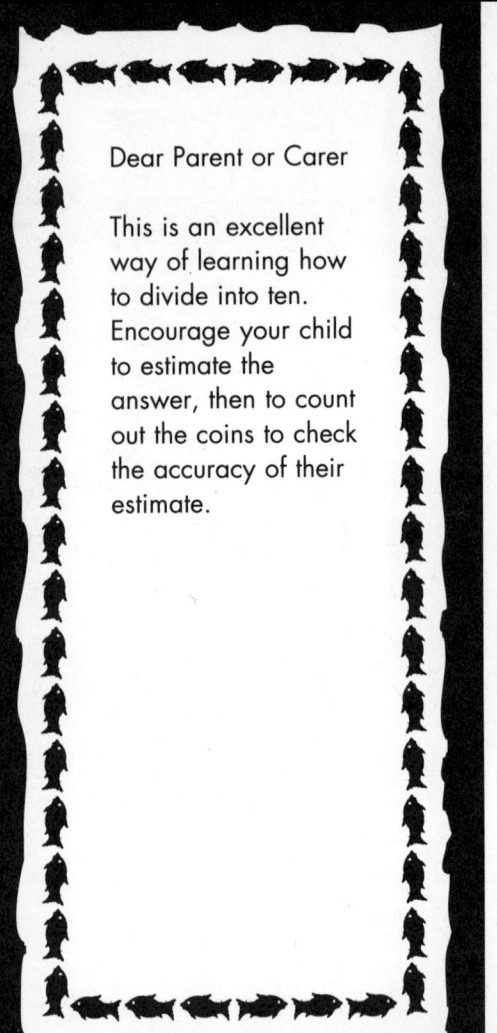

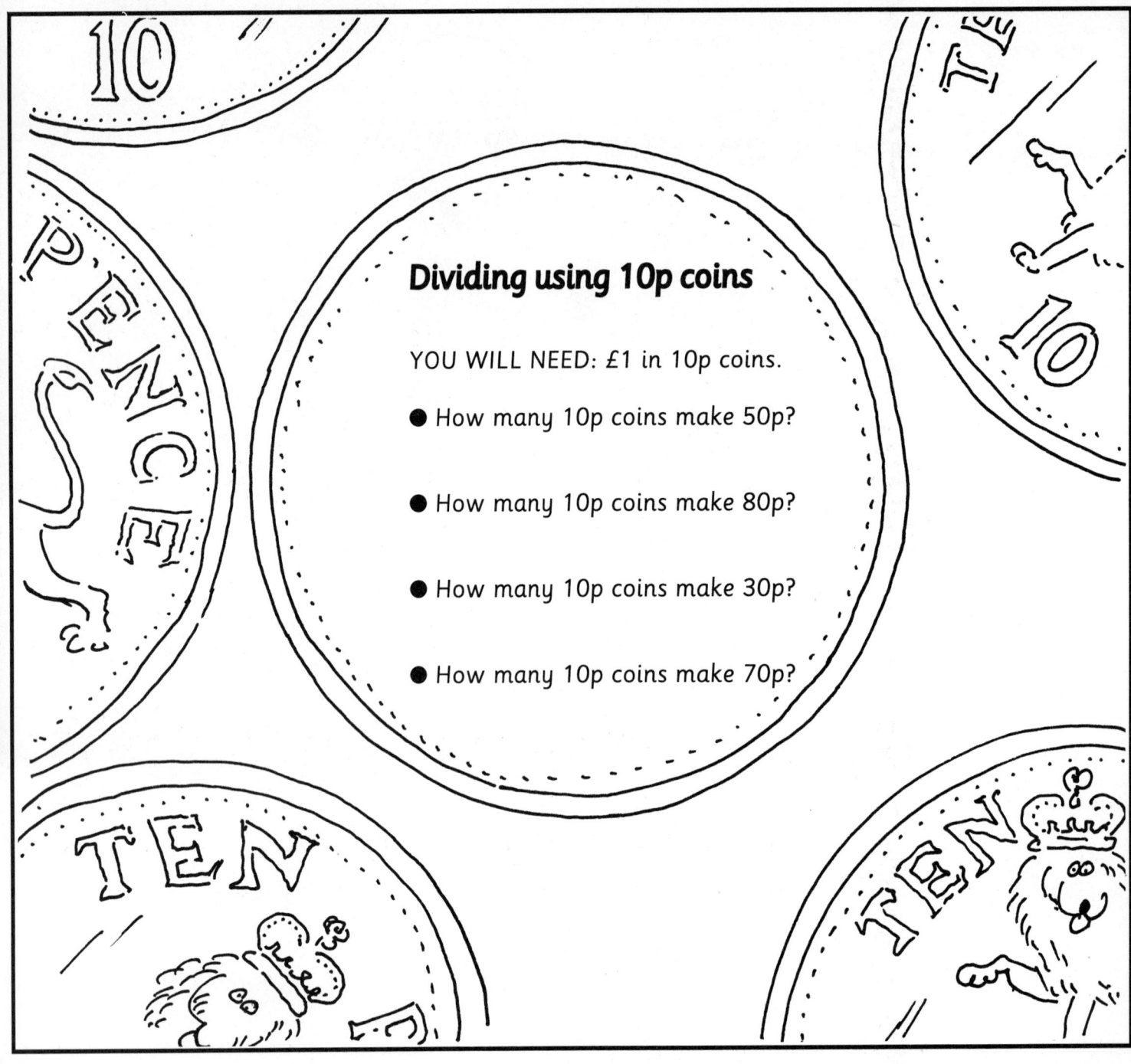

_____ and
child
_____
helper(s)
_____
did this activity together

48   **Holiday activities**

Dividing using 10p coins

YOU WILL NEED: £1 in 10p coins.

- How many 10p coins make 50p?

- How many 10p coins make 80p?

- How many 10p coins make 30p?

- How many 10p coins make 70p?

impact MATHS HOMEWORK

Adding tens and ones

YOU WILL NEED: plenty of 10p and 1p coins, some old Christmas or birthday cards and a pen.

- Make a set of cards numbered 0–9.

- Put two pairs of the number cards next to each other.

- Place coins of that value under each number; for example two 10p coins and six 1p coins.

- Now add both sets of numbers together. Remember to count the 10p coins first.

Dear Parent or Carer

Encourage your child to total the 10p coins first and tap each coin as it is counted: '10, 20, 30, 40, 50, 51, 52, 53, 54, 55, 56.' If the units total goes into a new ten number just continue counting. For example, 26 + 18: (that is, two 10p coins and six 1p coins and one 10p coins and eight 1p coins) '10, 20, 30, 31, 32, 33, 34, 35, 36, 37, 38, 39, 40, 41, 42, 43, 44.' Then compare this 44 with the simplest arrangement; that is, four 10p coins and four 1p coins.

_____ and
child

_____
helper(s)

_____
did this activity together

Holiday activities

Dear Parent or Carer

This is a fun game to increase your child's mental dexterity with numbers. If your child becomes very confident you could try reducing the numbers by 10 or increasing and/or decreasing by 100.

_____ and
child

_____
helper(s)

_____

did this activity together

50 **Holiday activities**

Ten more

This is a good game to play when travelling.

● Take turns to increase the number on the car in front by 10.

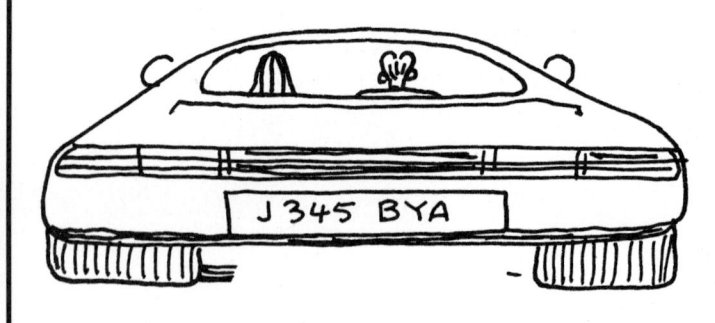

● Try the ones below first.

J 345 BYA

G 371 YAH

D 212 JBT

B 843 PMR

A 666 FDX

impact MATHS HOMEWORK

Can you multiply?

YOU WILL NEED: a pack of cards with the face (picture) cards removed (use the ace as 1), a few 2p, 5p and 10p coins, a beaker and someone to play with.

● Place the cards face down in a pile. Put the coins in a beaker.

● Take turns to take a card and a coin.

● Place the card first and say:

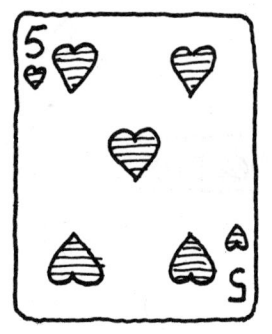

.... lots of .... makes ....
(card number)   (coin value)

Dear Parent or Carer

Your child may like to have several 10p, 2p and 5p coins so that each answer can be demonstrated, for example: 5 lots of 2p makes 10p and 2 lots of 5p makes 10p. (The answers are the same but the calculation is different.)

_____ and
child

_____
helper(s)

_____
did this activity together

Holiday activities

Dear Parent or Carer

Many children find calculating time difficult. Time can appear to go quickly or slowly depending on the activity. Please help your child to count in fives with the minute hand – it often helps to do this with a real clock. Then your child can see that the minute hand moves a lot faster than the hour hand.

_____ and
child

_____
helper(s)

did this activity together

TV planner

- Choose your favourite programmes from today's television schedule.

- Cut out this clock and hands and stick them on to some card (old birthday or Christmas cards).

- Use a split paper fastener to position the hands.

- Set the clock hands to the time that the earliest of your programmes starts. Mark this position with Blu-Tack.

- Move the hands to show when this programme ends. Count in fives to measure the length of the programme.

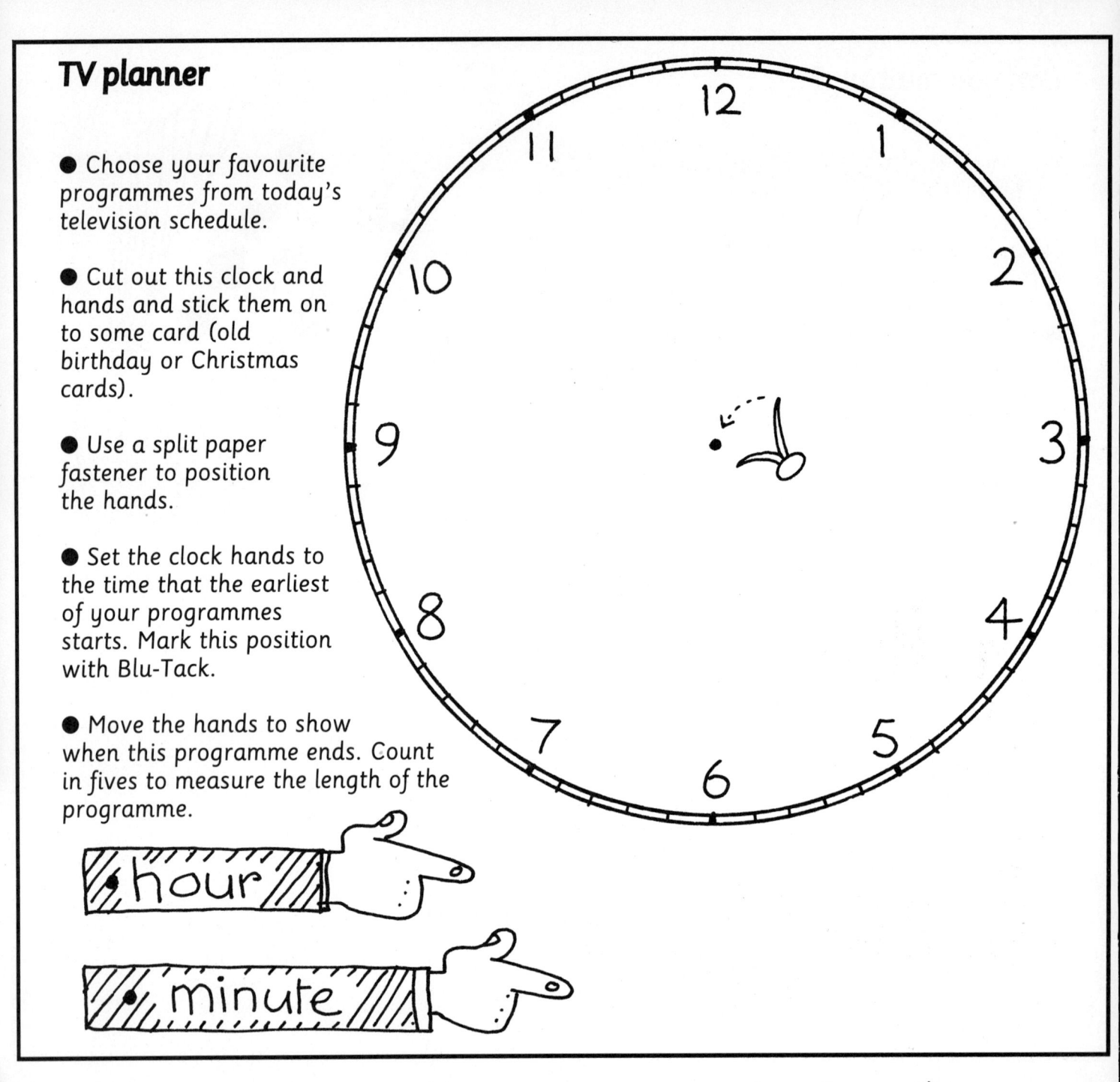

52 Holiday activities

impact MATHS HOMEWORK

Half

- Mark is half my age, he is .... years of age.

- How many sweets are in half this packet?

- Half my height is .... centimetres.

- It will be time for tea in half an hour. What time will it be?

- How many pages will you have read when you are halfway through this book?

- Half price sale; how much is teddy now?

£4·80

Dear Parent or Carer

Use this opportunity to discuss the various uses of half. Each time something is equally shared between two it is halved. Half the children sit on one side of the table, the other half sit on the other side. Half the sandwiches are egg and the other half are cheese and tomato, and so on.

_____and
child

_____
helper(s)

did this activity together

Holiday activities 53

impact MATHS HOMEWORK

Dear Parent or Carer

Many town centre church clocks show Roman numerals. Children are fascinated by anything different. You may like to arrange the numbers from 1–12 on a number line, ordinary numbers above and the corresponding Roman numerals below. Explain that IV is one before 5 and VI is one after 5.

_____ and
child

_____
helper(s)

_____
did this activity together

Holiday activities

Roman numbers

- Have you ever seen a clock or a watch with numbers like this?

- Can you work out what the numbers are? Write them beside the Roman numbers.

impact MATHS HOMEWORK

Where are we going?

YOU WILL NEED: a road map.

You are about to go on a car journey. Your helper will tell you where.

- Can you write down the roads and the places which you will travel through on your journey?

Dear Parent or Carer

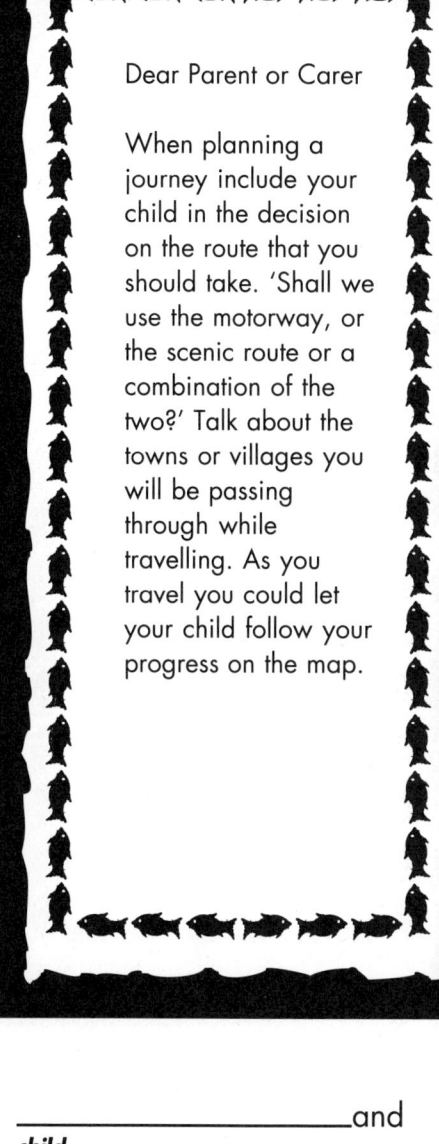

When planning a journey include your child in the decision on the route that you should take. 'Shall we use the motorway, or the scenic route or a combination of the two?' Talk about the towns or villages you will be passing through while travelling. As you travel you could let your child follow your progress on the map.

_____ and
child

_____
helper(s)

_____

did this activity together

Holiday activities

Shadows

Dear Parent or Carer

This activity could eventually lead on to observations about the rotation of the Earth on its axis. You may like to use this opportunity to discuss the position of the Sun in the sky: morning – east; midday – south; evening – west. You could also mention the fact that the Sun is always in the opposite direction to your shadow.

● Choose a sunny day. Ask somebody to measure your shadow length at least every two hours from the same position. They could use their feet to measure the length of your shadow.

● Does your height ever equal your shadow length? You can measure your height when you lie down – what are the differences?

● What do you notice? You may like to make a chart to record your findings, or draw a picture each time you look at your shadow. Remember to number your pictures to help you arrange them in order.

● Does this picture show a morning, lunch-time or evening shadow? How do you know?

_____and
child

_____
helper(s)

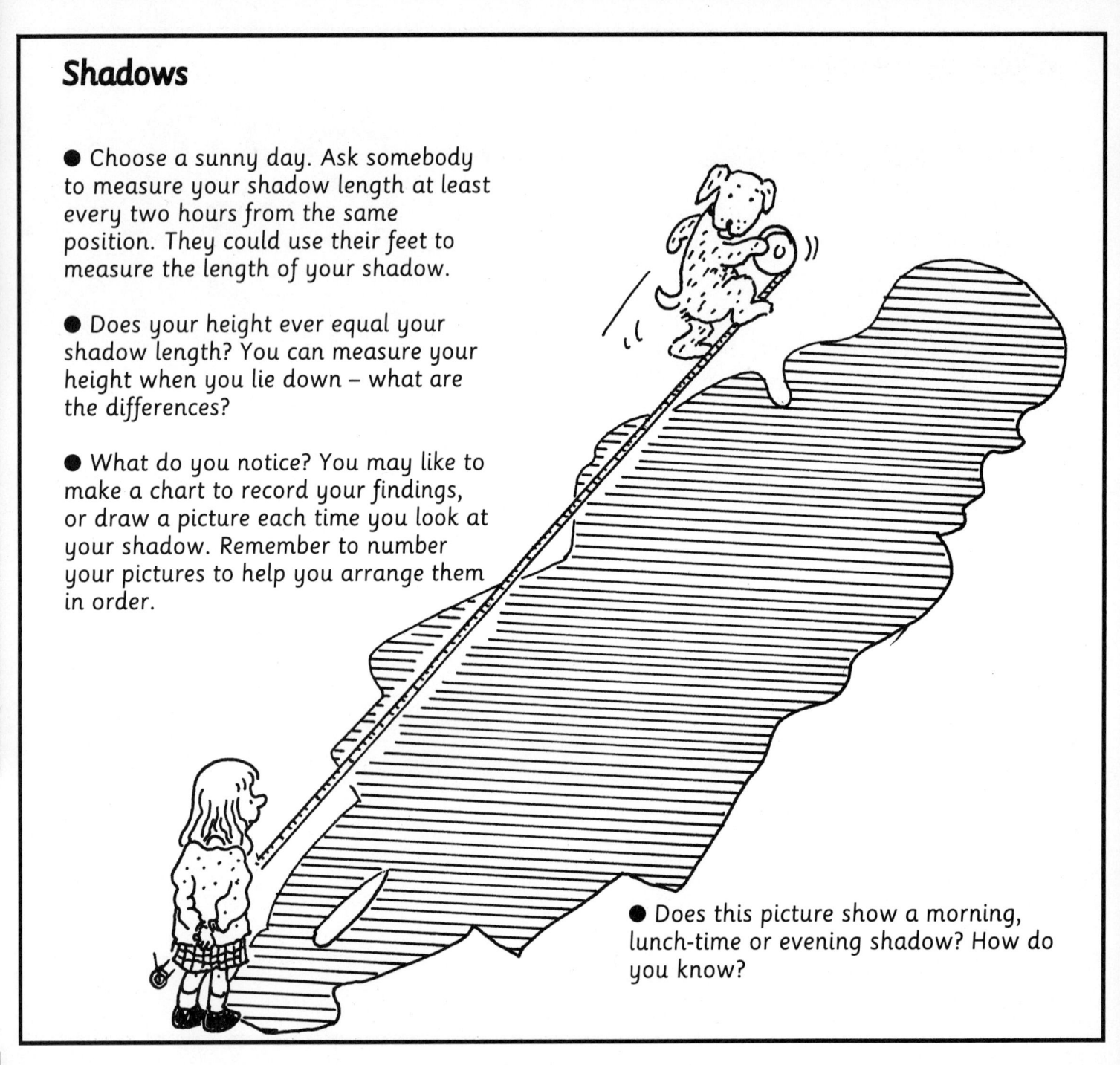

_____
did this activity together

Holiday activities

impact MATHS HOMEWORK

Make some two-digit numbers

YOU WILL NEED: some old birthday or Christmas cards numbered from 0–9.

- Make a number, for example:

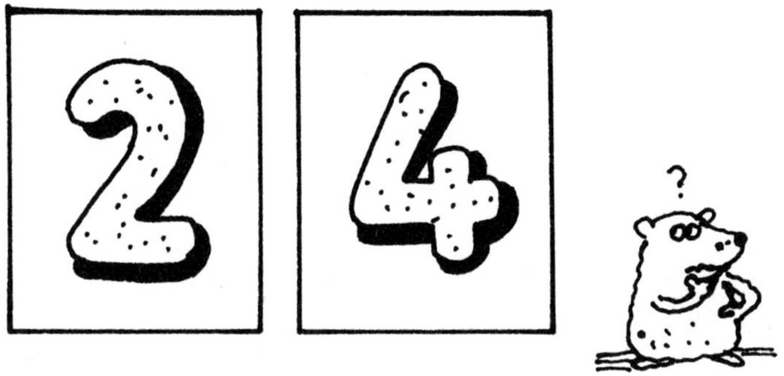

2 4

twenty four.

- Place another card under one of the numbers – what is the new number?

2 6 4

Is 26 a bigger or a smaller number than 24? How much bigger or smaller? Count on from 24: 25, 26 – it is 2 more.

- What is 2 more than (write the answer underneath):

30;   43;   15;   27;   56;   78?

If a number is smaller than 24, for example, 22, you count back from 24: 23, 22 – it is 2 less.

- What is 2 less than (write the answer underneath):

27;   36;   19;   48;   54;   73?

Dear Parent or Carer

This game helps your child to realise the significance of tens and units (ones) in our number system. Most children find it easier to increase numbers than to decrease.

_____and
child

_____
helper(s)

_____

did this activity together

Holiday activities

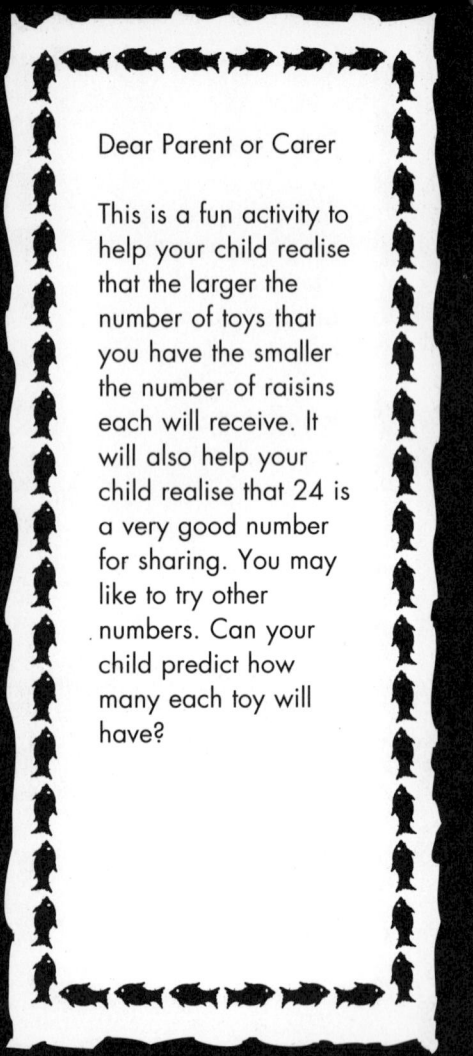

Dear Parent or Carer

This is a fun activity to help your child realise that the larger the number of toys that you have the smaller the number of raisins each will receive. It will also help your child realise that 24 is a very good number for sharing. You may like to try other numbers. Can your child predict how many each toy will have?

_____and
child

_____
helper(s)

_____
did this activity together

58 **Holiday activities**

Sharing

YOU WILL NEED: 24 objects (LEGO, 1p coins or raisins) and several toys.

● Begin with two toys. Share the raisins between the toys – how many will they each receive?

● Put the raisins back in the middle and try sharing between three toys, four toys, five toys and so on.

impact MATHS HOMEWORK

Coin time-line

YOU WILL NEED: a handful of coins.

● Look at the last two digits of the date on each coin, for example 19<u>84</u> and arrange your coins in age order.

Dear Parent or Carer

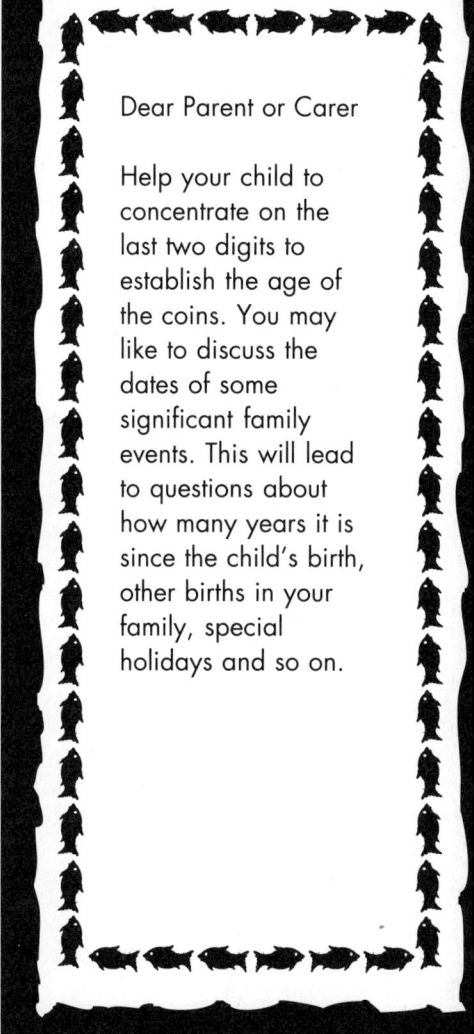

Help your child to concentrate on the last two digits to establish the age of the coins. You may like to discuss the dates of some significant family events. This will lead to questions about how many years it is since the child's birth, other births in your family, special holidays and so on.

_____ and
child

_____
helper(s)

_____
did this activity together

Holiday activities 59

Dear Parent or Carer

Nought is an extremely important number as it sets the place value in our number system. It is often easier to ask children to read numbers like 240 by beginning at the right-hand side; for example, first you read 0, then 40 and then 240.

_____and
child

_____
helper(s)

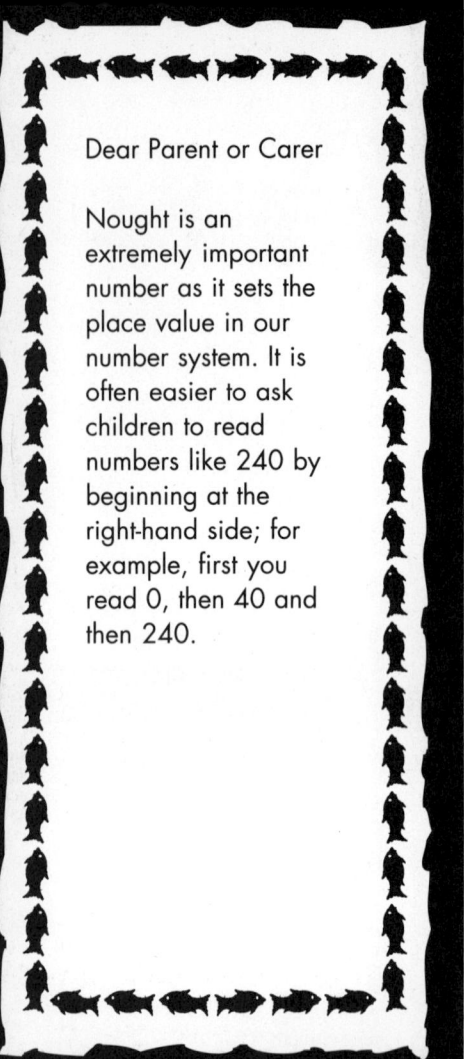

_____
did this activity together

60 Holiday activities

Using 0 to make big numbers

YOU WILL NEED: old birthday or Christmas cards numbered 0–9.

● Take the 0 card and any two others. How many numbers can you make?

● Can you read all the numbers and write them in order?

impact MATHS HOMEWORK

Reading car numbers

Take turns to be first at playing this game on a car journey.

- The first player reads the numbers on the car in front, for example,

F 357 LBW

357 'Three hundred and fifty seven.'

- The second player reads the next car number that comes in front, for example,

K294 KBT

294 'Two hundred and ninety four.'

- The first player gets 1 point because his number is bigger, the winner of the game is the player who reaches 5 points first.

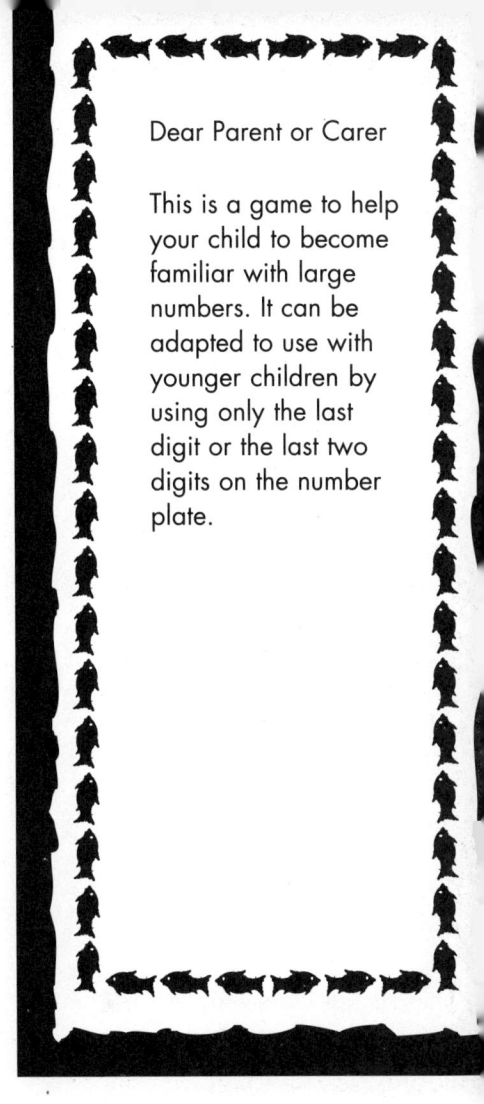

Dear Parent or Carer

This is a game to help your child to become familiar with large numbers. It can be adapted to use with younger children by using only the last digit or the last two digits on the number plate.

_____ and
child

_____
helper(s)

_____
did this activity together

Holiday activities

Dear Parent or Carer

This is an excellent game for helping your child to understand how numbers become larger or smaller by changing their position (place value). You could take turns to make up the biggest or the smallest number.

_____and
child

_____
helper(s)

_____
did this activity together

Arranging car numbers

● Look at the number plates on the cars below.

● Can you rearrange the numbers to make a bigger number? On the first car the biggest number you can make is six hundred and thirty two.

● What is the smallest number which can be made with this plate?

● Try out this game on a car journey.

Holiday activities

50% off!

Everything is at half price in the toy shop today.

- Price your toys and arrange them in a shop.
- Borrow some coins from somebody at home, so that you can play shops with a friend.
- You will need to be able to give your customers the new prices when they pay for your toys.
- Practise on the examples below first.

£1.30   £8.44   £3.28   £2.24

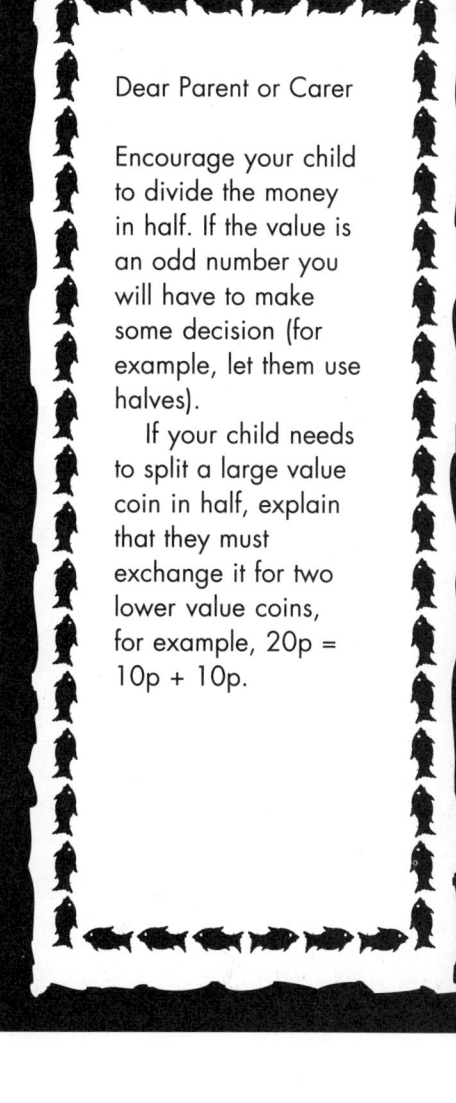

Dear Parent or Carer

Encourage your child to divide the money in half. If the value is an odd number you will have to make some decision (for example, let them use halves).

If your child needs to split a large value coin in half, explain that they must exchange it for two lower value coins, for example, 20p = 10p + 10p.

_____ and
child

_____
helper(s)

_____
did this activity together

Holiday activities

Dear Parent or Carer

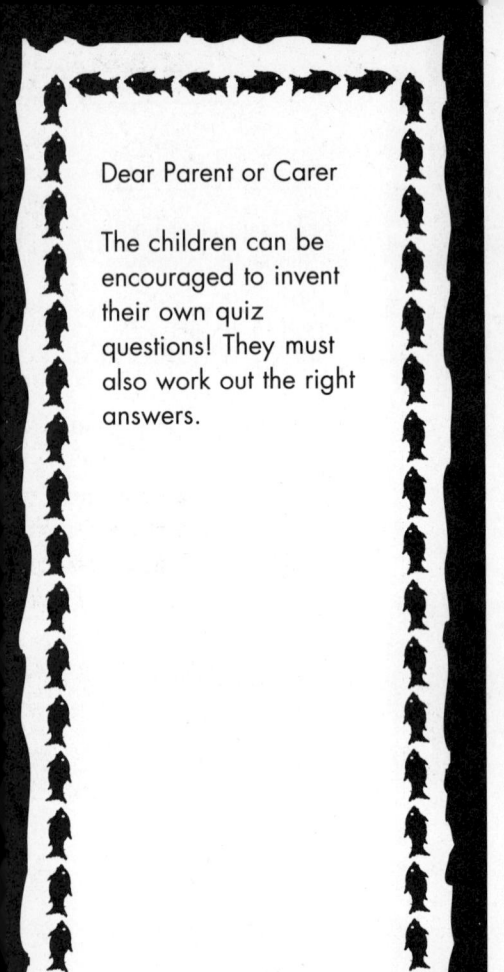

The children can be encouraged to invent their own quiz questions! They must also work out the right answers.

_____ and
child

_____
helper(s)

_____
did this activity together

Holiday activities

Toy quiz

● Line up as many of your soft toys as you can. You are going to give them a 'teddy quiz'!

● Cut out the questions along the bottom of the page and give them out. Help each of your toys to write in the answers.

● Ask a grown-up to check their work with you. Which toy gets the most right answers?

I am a two-figure number whose digits add up to 18. Which number am I?	I am in the 2× table and in the 3× table and in the 5× table. What number am I?	I am a quarter of twice 4 doubled. What number am I?
I am 3 larger than half of 50. What number am I?	I am the only even prime number. What number am I?	I am 10 times 10 times 10. What number am I?
I am one less than 1000. What number am I?	I am one more than 1000. What number am I?	I am a three-figure number and my digits add up to 3. What number am I?

impact MATHS HOMEWORK

Snakey numbers!

- Write all the numbers you can on to the snake in the right order, starting at his head.

- Before you do it, can you guess what number will wind up at his tail?

- Ask as many people as you can to guess. Then write in all the numbers and find out who was right.

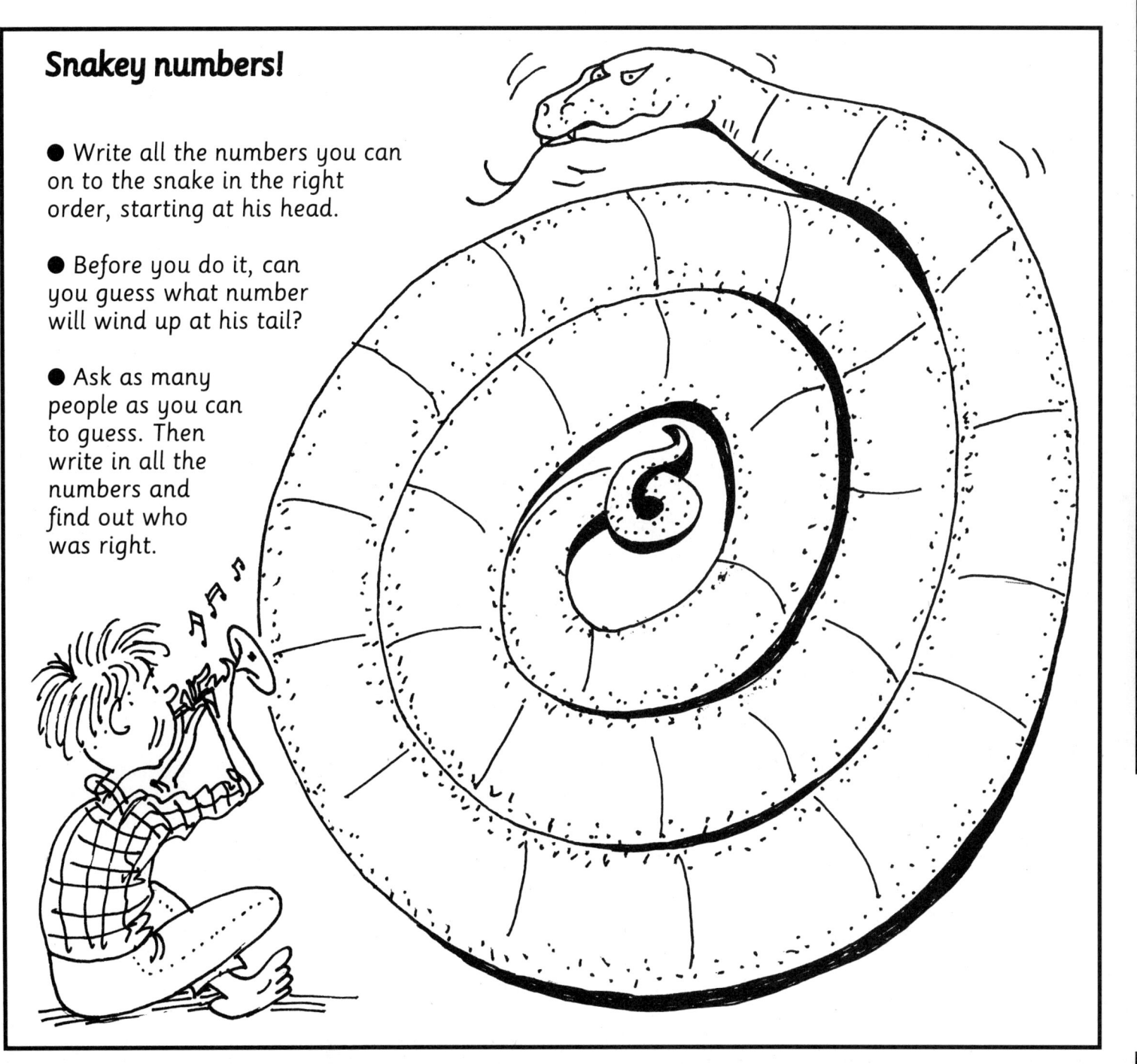

Dear Parent or Carer

It is surprising the number of people who guess wrong here! Help your child if they get stuck with any of the numbers.

_____ and

child

_____

helper(s)

_____

did this activity together

Holiday activities

Dear Parent or Carer

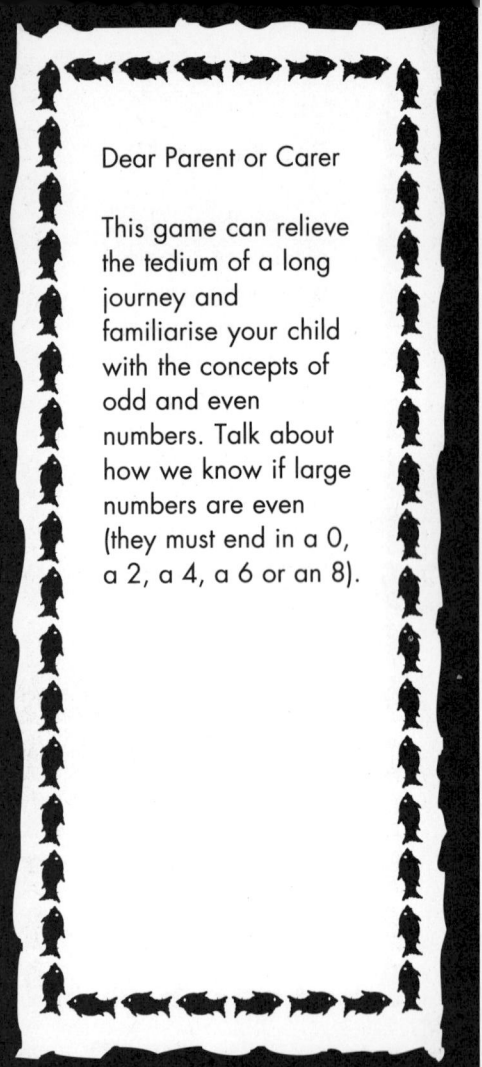

This game can relieve the tedium of a long journey and familiarise your child with the concepts of odd and even numbers. Talk about how we know if large numbers are even (they must end in a 0, a 2, a 4, a 6 or an 8).

_____ and
child

_____
helper(s)

_____
did this activity together

Holiday activities

Car cricket

● Play the following car game on long boring car journeys.

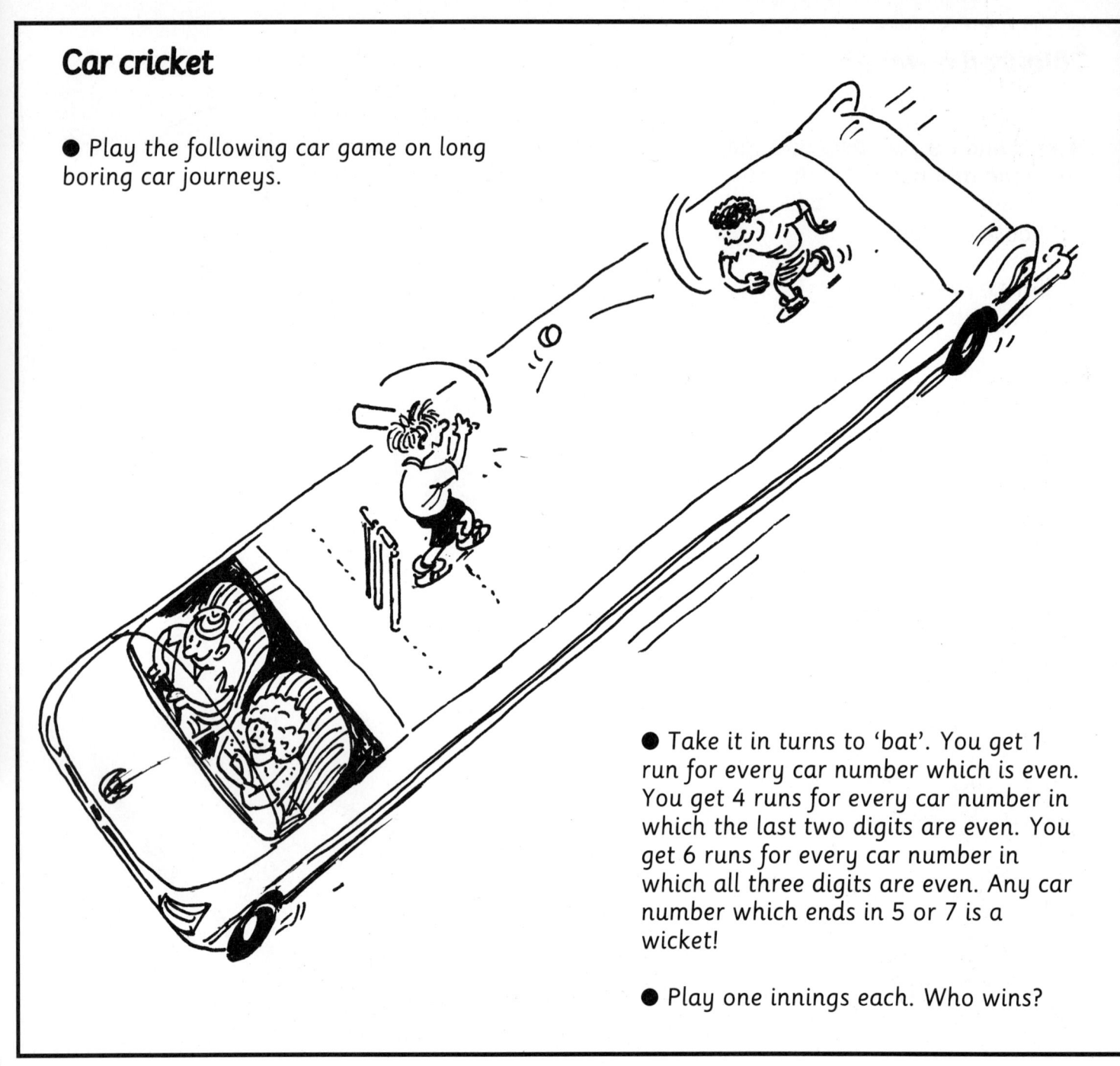

● Take it in turns to 'bat'. You get 1 run for every car number which is even. You get 4 runs for every car number in which the last two digits are even. You get 6 runs for every car number in which all three digits are even. Any car number which ends in 5 or 7 is a wicket!

● Play one innings each. Who wins?

impact MATHS HOMEWORK

Coloured numbers

Here is one way to amuse yourselves on a long and boring car journey!

- Choose a colour each.
- Choose even or odd numbers.
- If you choose even and you see a car of your colour with an even number, score 10 points.

  If you choose odd and you see a car of your colour with an odd number, score 10 points.

  If you choose even, and you see a car of your colour with an even number which is also even when halved, score 20 points. (For example, 248 is 124 when halved which is even.)

  If you choose odd, and you see a car of your colour with a number which is also odd, but if you add 1 and halve it to make an even number, score 20 points. (For example, 251 is 252 when you add 1 and when halved makes 126 which is even.)

- The first person to score 100 points is the winner!

impact MATHS HOMEWORK

Dear Parent or Carer

The children will do quite a lot of mental arithmetic when playing this game. Encourage them to say the numbers out loud – not as 'three, five, two' for 352, but as 'three hundred and fifty-two' – this helps them with reading large numbers.

_____and

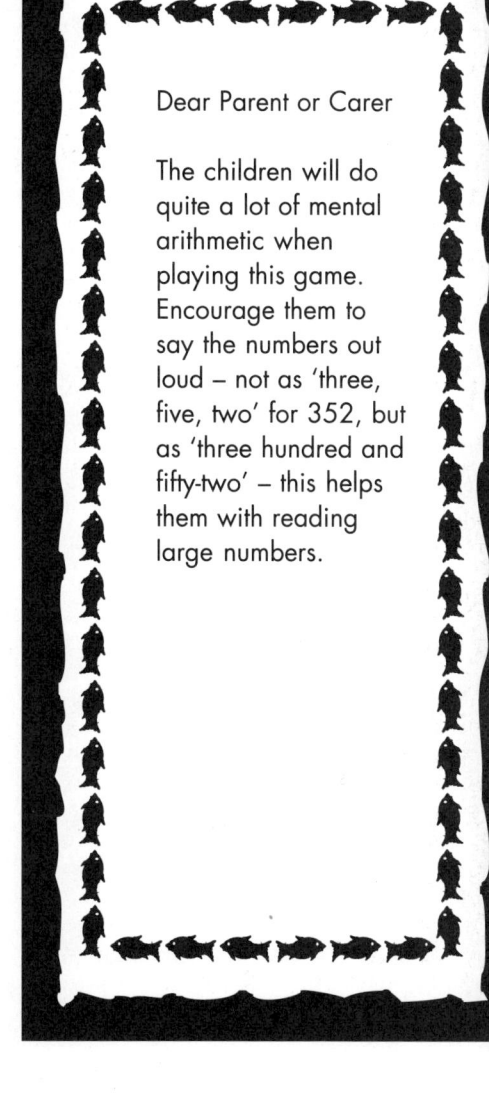

child

_____

helper(s)

_____

did this activity together

Holiday activities 67

Dear Parent or Carer

This activity not only helps children practise counting skills – it is hard to count so many pieces accurately! – it also helps them to organise a data-collecting sheet when collecting the guesses.

_____and
child

_____
helper(s)

_____
did this activity together

Guess-timate

YOU WILL NEED: a mug filled with rice.

● Ask everyone you can (including yourself) to guess how many pieces there are in the mug!

● Write down their names with their guesses beneath.

● Now you have to count the rice pieces! Do this by putting them in piles of 10 or 20 or even 25.

● Who was closest?

Holiday activities

Car search

Play this game with someone on a boring car journey!

- Agree a target number – it must be three figures, for example, 357.

- Start looking for cars which have the same last two digits, for example 857.

- If you find one, score the difference between your number and the target number.

- Play for as long as the journey or till you get bored! Who wins?

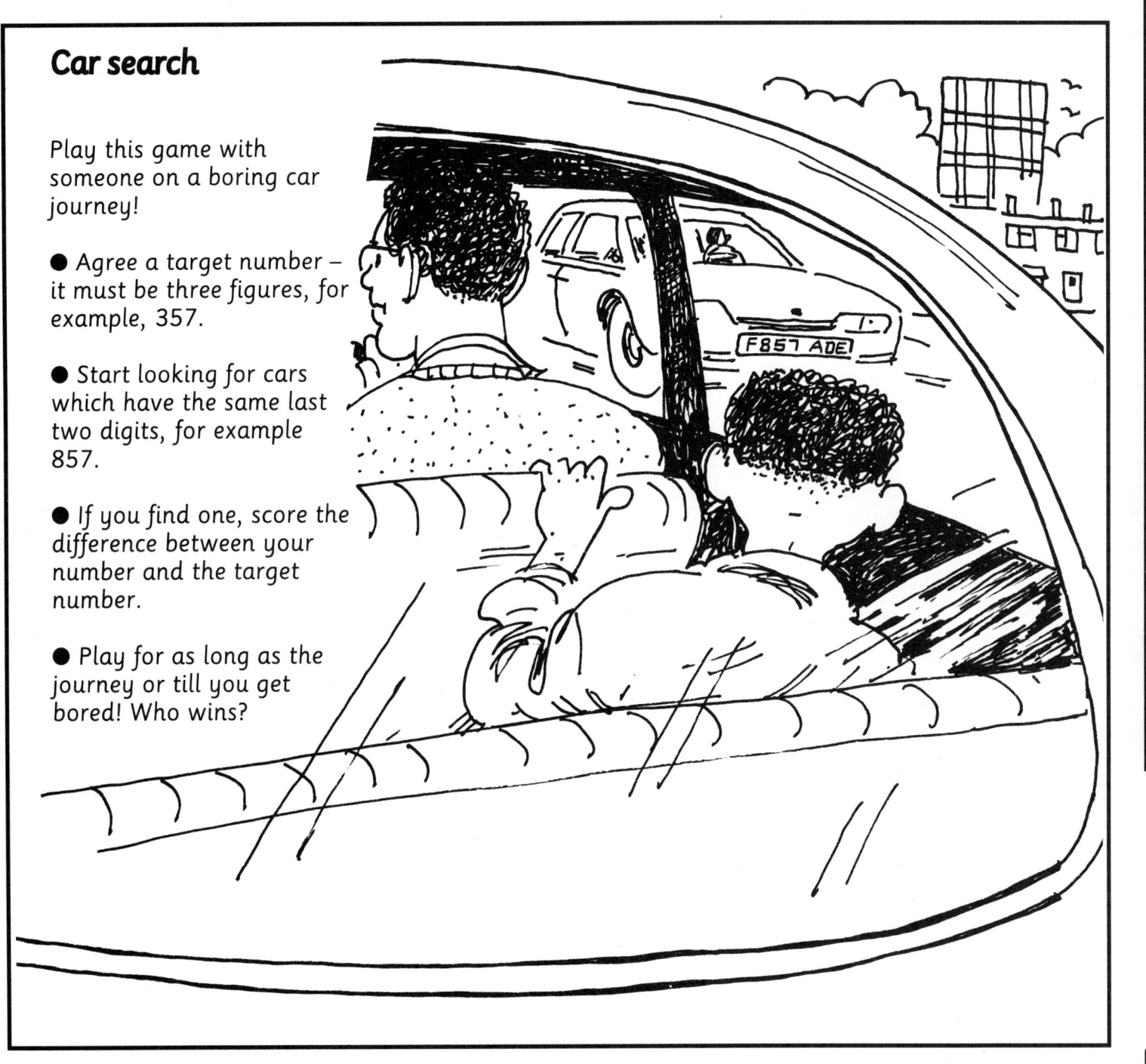

impact MATHS HOMEWORK

Dear Parent or Carer

This activity is excellent for mental arithmetic. You will have to help your child work out the differences – it is not at all easy!

_____ and
child

_____
helper(s)

_____

did this activity together

Holiday activities 69

Dear Parent or Carer

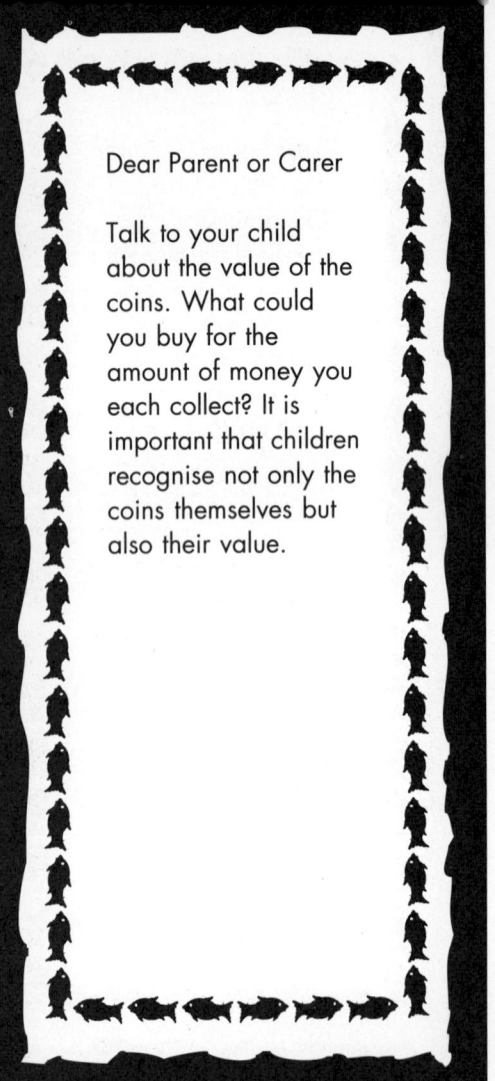

Talk to your child about the value of the coins. What could you buy for the amount of money you each collect? It is important that children recognise not only the coins themselves but also their value.

_____ and
child
_____
helper(s)
_____
did this activity together

Holiday activities

Throwing chances!

YOU WILL NEED: a handful of coins and someone to play with.

● Spread out the coins. Choose one of you as 'heads' and the other as 'tails'.

● Take it in turns to choose a coin and spin it. If it lands your way (heads or tails, depending on what you chose), keep the coin.

● Keep playing until all the coins are taken!

● Add up your coins. Who has the most? Play again, changing tails and heads. Who wins this time?

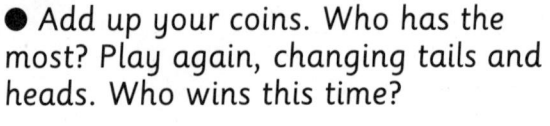

impact MATHS HOMEWORK

Weight guesses

- Can you find some things which you think will weigh exactly half a kilogram (that is, 500 grams).

- Collect toys, or LEGO bricks, or anything else you can and place them in a small bag.

- Now check their weight. How close were you?

- Can someone else in your home find some things that they think will together weigh 500 grams?

(Remember, no cheating by using packets of food which are marked with the weight!)

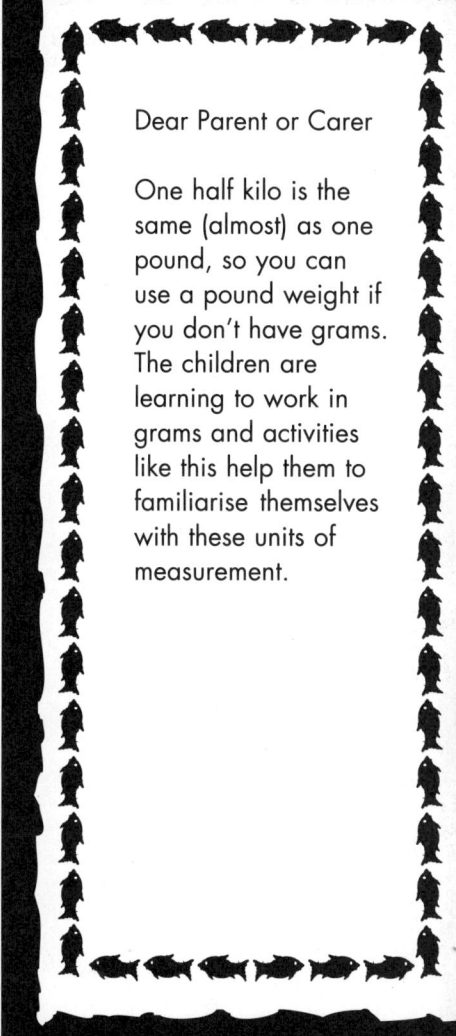

impact MATHS HOMEWORK

Dear Parent or Carer

One half kilo is the same (almost) as one pound, so you can use a pound weight if you don't have grams. The children are learning to work in grams and activities like this help them to familiarise themselves with these units of measurement.

_____ and
child

_____
helper(s)

_____
did this activity together

Holiday activities 71

Dear Parent or Carer

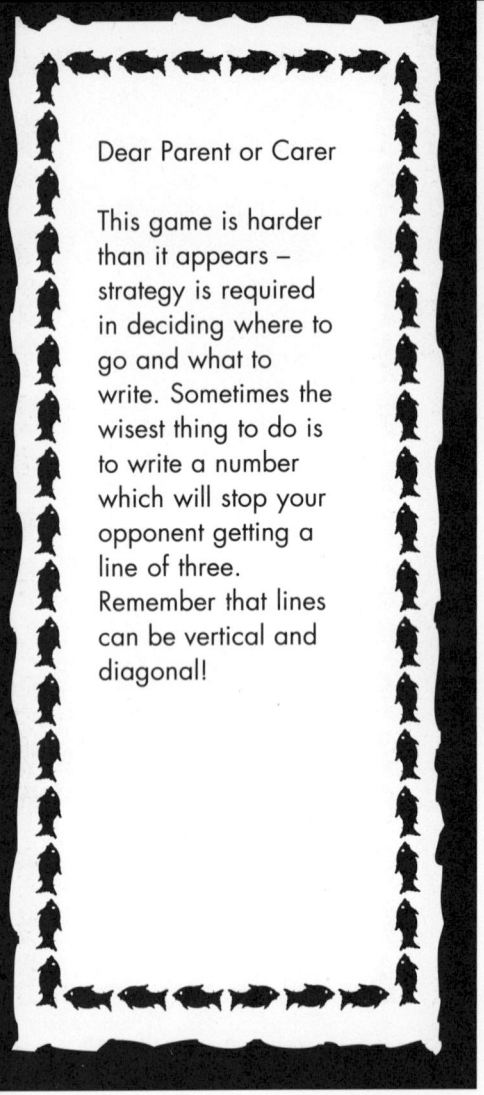

This game is harder than it appears – strategy is required in deciding where to go and what to write. Sometimes the wisest thing to do is to write a number which will stop your opponent getting a line of three. Remember that lines can be vertical and diagonal!

_____ and
child

_____
helper(s)

_____

did this activity together

Three in a line

YOU WILL NEED: a pencil each and someone to play with.

- Using one of the grids below, take it in turns to write numbers in the squares. You may write any number between 0 and 9 inclusive in any square on the grid.

- If you write a number that makes a straight line of three numbers in any direction which adds up to 10, you may score 10 and put a line through those three numbers. None of those numbers may be used again in another line.

- When all the squares are full of numbers, the person with the highest score wins.

Holiday activities

Hidden code

The puzzle below has a hidden code number.
● Colour in all the spaces marked with multiples of 9 to find out what the code number is.

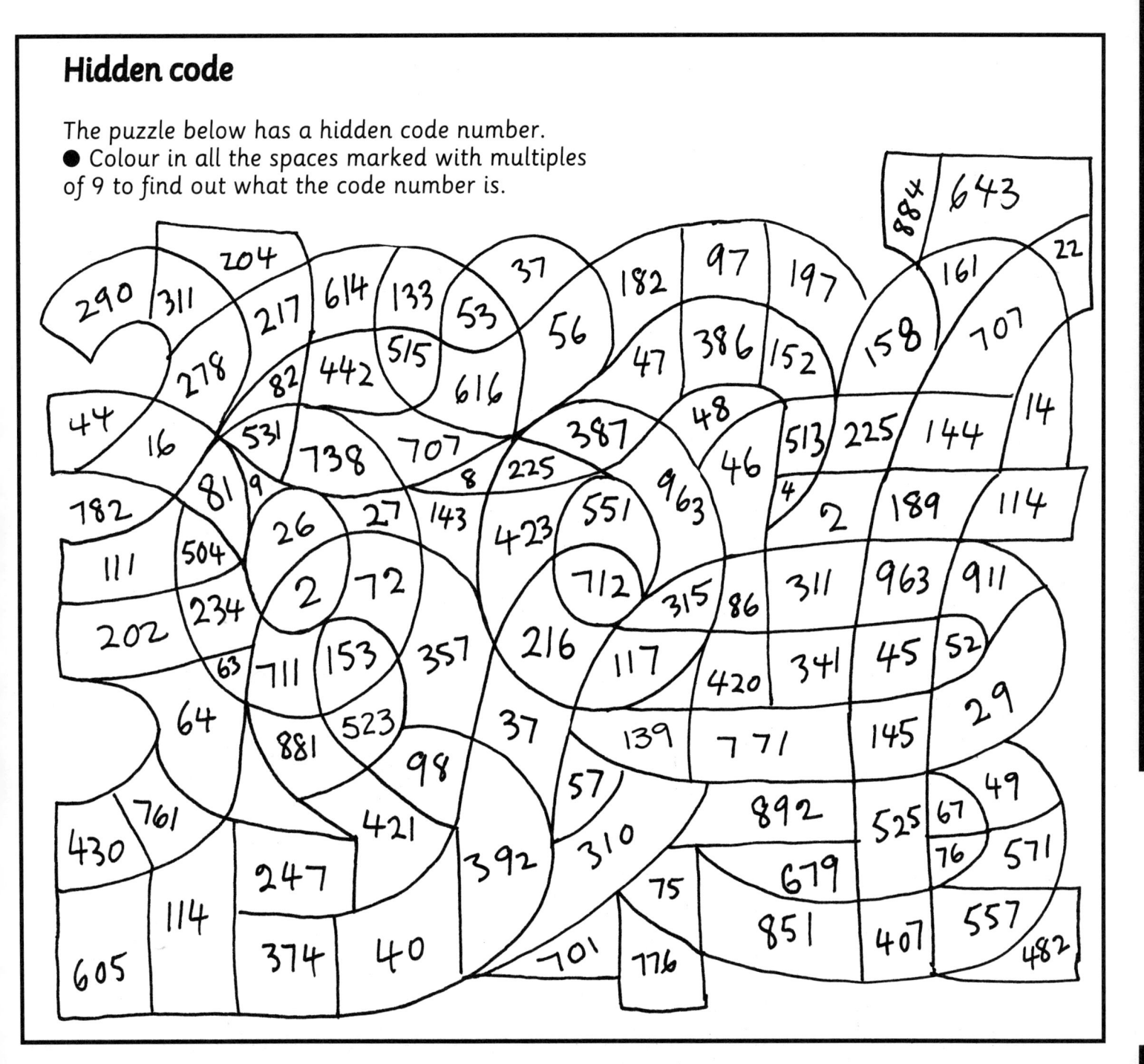

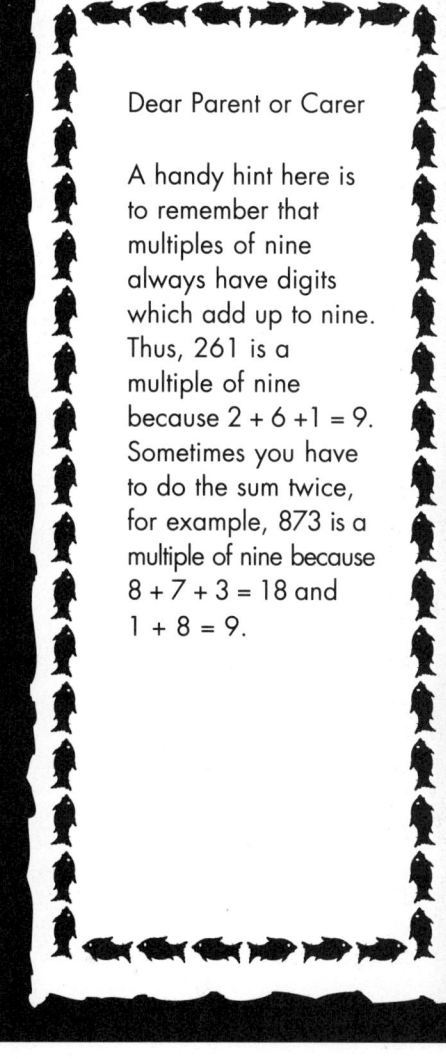

Dear Parent or Carer

A handy hint here is to remember that multiples of nine always have digits which add up to nine. Thus, 261 is a multiple of nine because 2 + 6 + 1 = 9. Sometimes you have to do the sum twice, for example, 873 is a multiple of nine because 8 + 7 + 3 = 18 and 1 + 8 = 9.

_____ and
child

_____
helper(s)

_____

did this activity together

Holiday activities

Dear Parent or Carer

This game is very good for children's mental arithmetic. Make it as much fun as possible so that they keep trying to score goals!

_____ and
child

_____
helper(s)

_____
did this activity together

Holiday activities

Car football

Play this version of car football on a long and boring journey.

- Each choose a team you want to be.

- One team looks for cars whose number plates have numbers which add up to 10. Every time they find one they score a goal.

- The other team looks for cars whose number plates add up to 11. Every time they find one they score a goal.

- The winner is the first team to get 10 goals.

impact MATHS HOMEWORK

Pasta count-up

YOU WILL NEED: a packet of pasta shapes.

- Can you guess how many pieces of pasta there are in a packet?

- Ask as many people as you can find to guess too.

- Measure 100 grams in a weighing machine.

- Count how many pieces there are.

- Now work out how many lots of 100 grams there are in the packet.

- What calculation will you now have to make to find out how many pieces there are altogether in the packet?

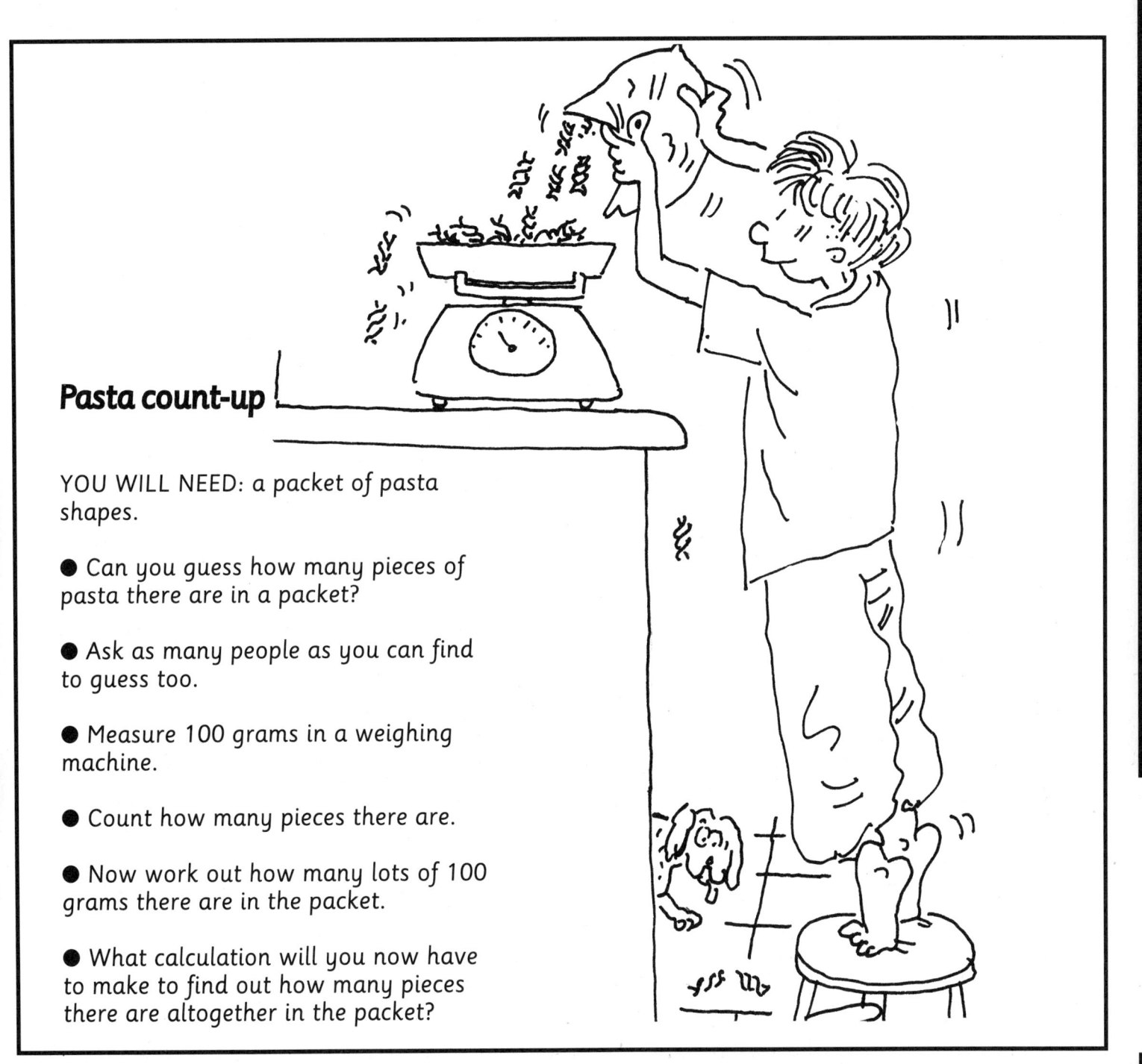

Dear Parent or Carer

Help the child to do the necessary calculations if they find them hard. It may be helpful to use a calculator.

_____and
child

_____
helper(s)

_____

did this activity together

Holiday activities

Dear Parent or Carer

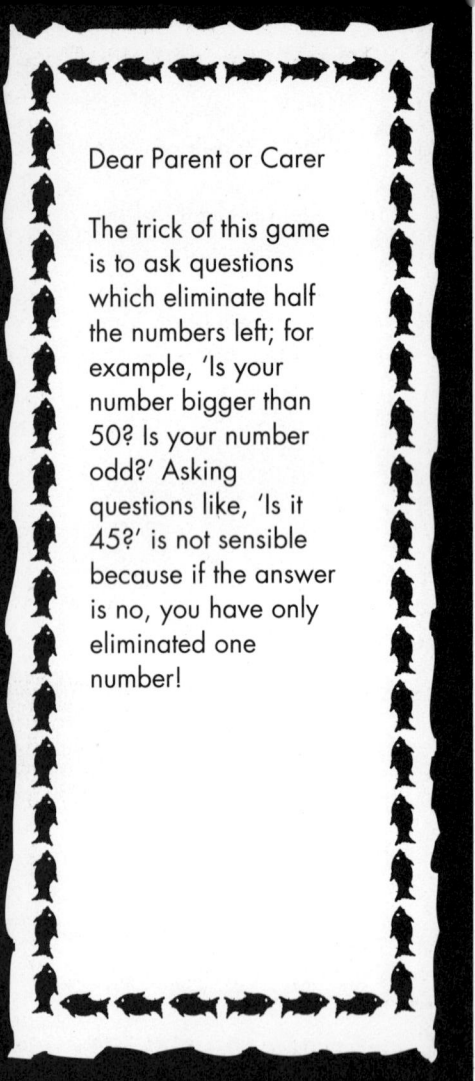

The trick of this game is to ask questions which eliminate half the numbers left; for example, 'Is your number bigger than 50? Is your number odd?' Asking questions like, 'Is it 45?' is not sensible because if the answer is no, you have only eliminated one number!

_____ and
child

_____
helper(s)

_____
did this activity together

Bluff it out!

This game relies upon keeping a straight face!

- Think of a number between 1 and 100.

- Then someone has to ask you six questions about your number.

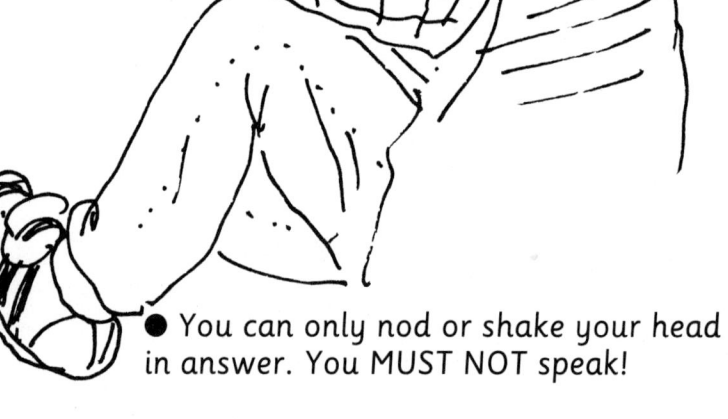

- You can only nod or shake your head in answer. You MUST NOT speak!

- Do they guess your number in six questions?

- Now they think of a number and you guess theirs. Do you succeed?

Holiday activities

impact MATHS HOMEWORK

Harder car cricket

Play the following car game on long boring journeys.

- Take it in turns to 'bat'.
- You get 1 run for every car number which divides by 2.
- You get 4 runs for every car number which divides by 4 (that is, you halve it and it is still even).
- You get 6 runs for every car number which divides by 6 (that is, is even and divides by 3).
- Any car number which ends in 5 or 7 is a wicket!
- Play one innings each. Who wins?

Dear Parent or Carer

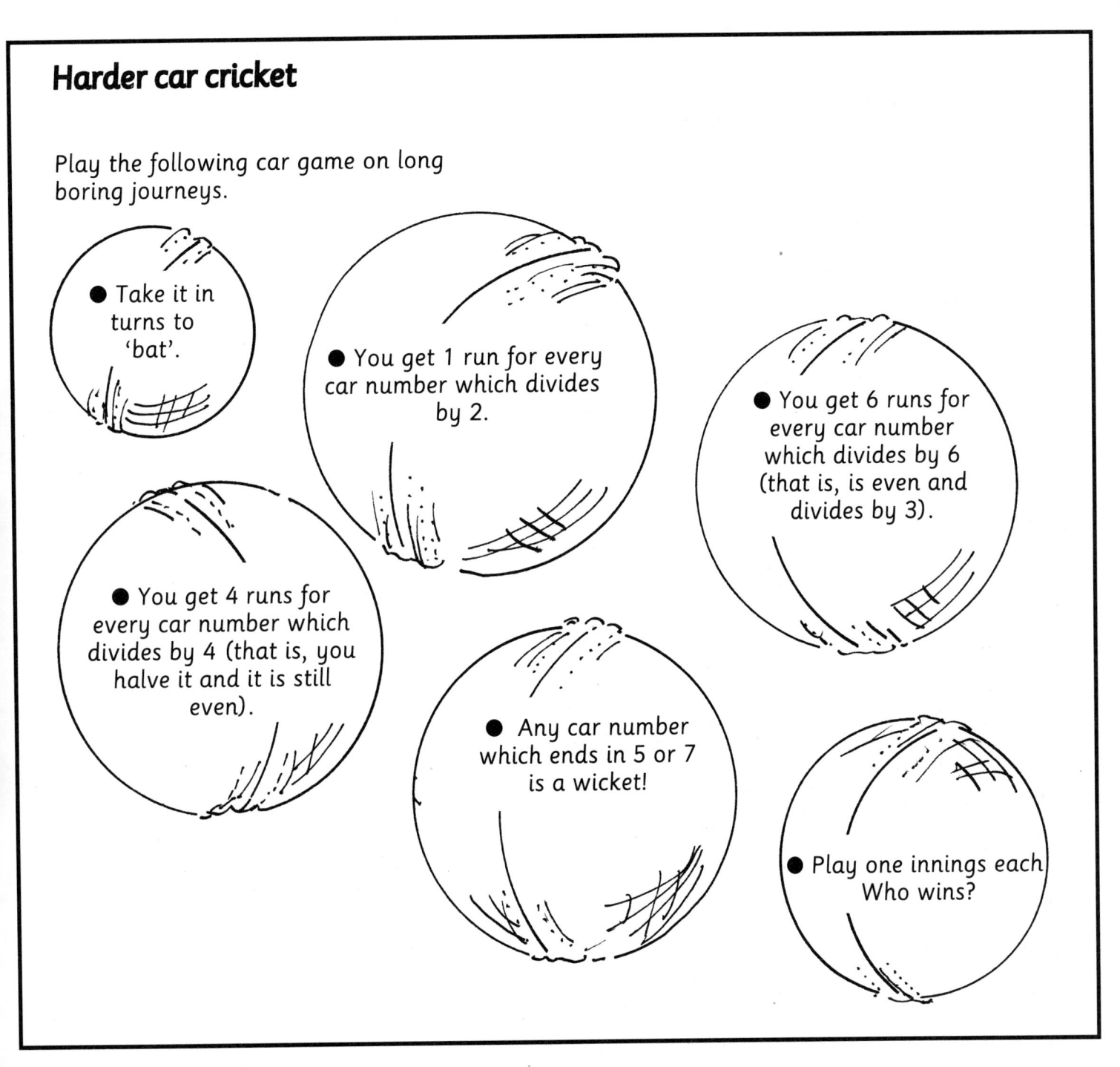

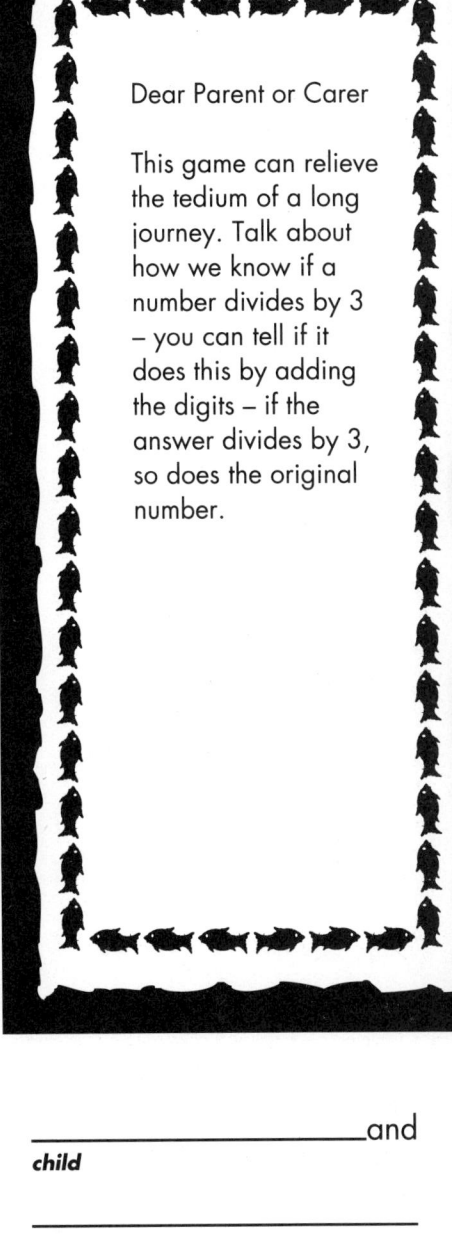

This game can relieve the tedium of a long journey. Talk about how we know if a number divides by 3 – you can tell if it does this by adding the digits – if the answer divides by 3, so does the original number.

_____ and *child*

_____ *helper(s)*

_____

did this activity together

Holiday activities

Dear Parent or Carer

This activity is excellent for mental arithmetic. You will have to help the children work out the differences – it is not at all easy!

_____ and
child

_____
helper(s)

_____
did this activity together

78   **Holiday activities**

Number search

Play this game with someone on a boring car journey!

- Agree a target number – it must be three figures – for example, 357.

- Start looking for cars which have these three digits in any order, for example, 735.

- If you find one, score the difference between your number and the target number.

- Play for as long as the journey or till you get bored! Who wins?

impact MATHS HOMEWORK

Ten pence

- Try to convince someone you know to give you as many 10p coins as there are different ways of making ten pence using different combinations of coins!

- How much do you – and they – think they will have to give you?

- Now work out how many ways there are to make ten pence. They must all be different!

- Write them all down in a list. How do you know you have found them all?

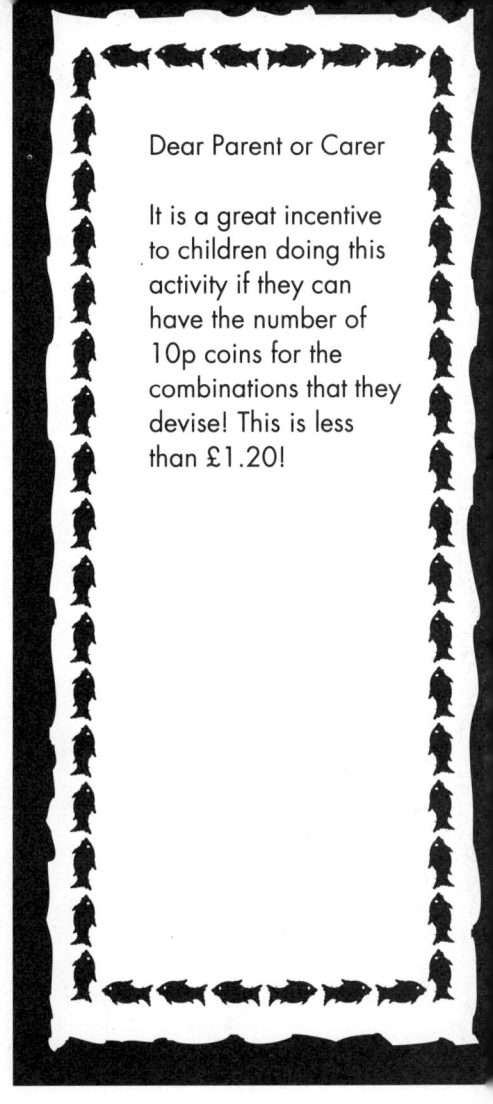

Dear Parent or Carer

It is a great incentive to children doing this activity if they can have the number of 10p coins for the combinations that they devise! This is less than £1.20!

_____ and
child

_____
helper(s)

_____
did this activity together

Holiday activities

impact MATHS HOMEWORK

Dear Parent or Carer

Games are a really good way to practise arithmetic and especially the times tables.

_____and
child
_____
helper(s)
_____
did this activity together

Holiday activities

Card tables

YOU WILL NEED: a pack of cards.

- Remove the face (picture) cards. Deal out five cards to each player.

- Each person is trying to make 24 using some or all of the cards in their hand. You can do this any way you like. For example:

$(4 + 2) \times 4 = 24$
or $8 \times (1 + 2) = 24$
or $10 \times 3 - 6 = 24$

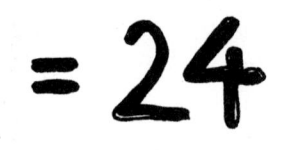

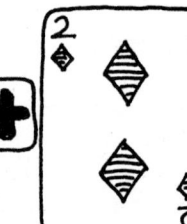

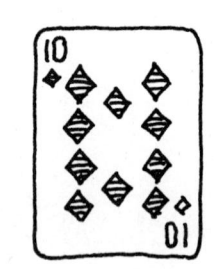

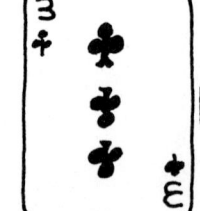

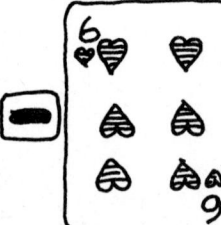

- Every time you make 24, lay the cards down in front of you. Replace them by taking some more from the pack. If you can't make 24 with the cards in your hand, take another card from the pack.

- Play until all the cards are gone! The winner is the person who made the most 24s.

4 + 2 × 4 = 24

8 × A + 2 = 24

10 × 3 − 6 = 24

and so on.

impact MATHS HOMEWORK

Scale animal

- In the small grid below, carefully draw a picture of your favourite animal and colour it in.

- Then copy the picture on to the larger grid. Do this by copying **exactly** what is in each square on the smaller grid on to each square on the larger grid.

- Colour it in. Compare your two pictures. The larger one is nine times the size of the smaller one!

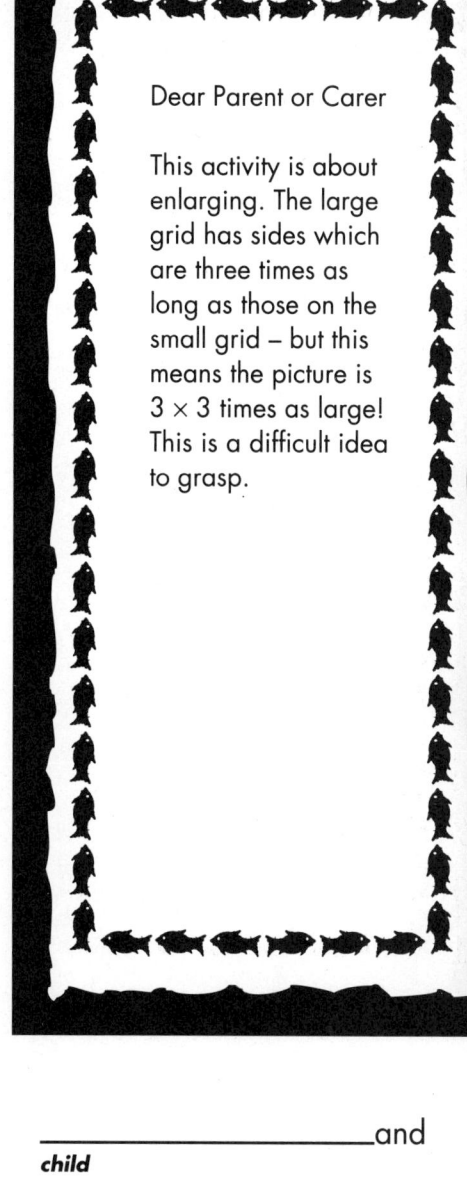

Dear Parent or Carer

This activity is about enlarging. The large grid has sides which are three times as long as those on the small grid – but this means the picture is 3×3 times as large! This is a difficult idea to grasp.

_____ and
child

_____
helper(s)

_____
did this activity together

Holiday activities 81

impact MATHS HOMEWORK

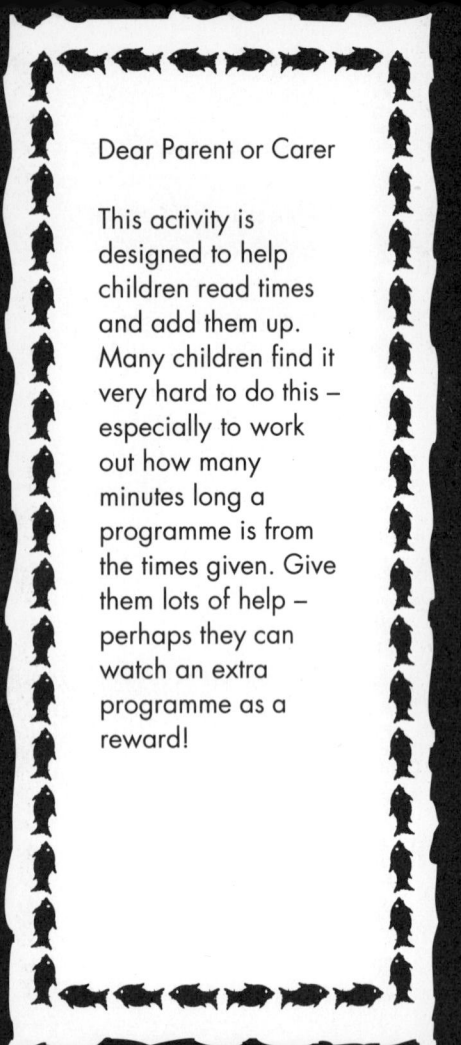

Dear Parent or Carer

This activity is designed to help children read times and add them up. Many children find it very hard to do this – especially to work out how many minutes long a programme is from the times given. Give them lots of help – perhaps they can watch an extra programme as a reward!

_____ and
child

_____
helper(s)

_____
did this activity together

82 **Holiday activities**

TV marathon

- Suppose that, in a moment of wild abandon, your parents agreed to let you watch 24 hours TV in two days!

- Which programmes would you watch?

- Using the TV page, plan your 24 hours viewing! Carefully plan which programmes on each channel you would watch. Keep a total of the number of hours viewing because it must total exactly 24 hours!

- Make a list and show it to a grown-up in your house. Which programmes **will** they let you watch?

impact MATHS HOMEWORK

Unusual view

Have you ever seen those camera shots of an ordinary household object which make it look like something really weird?!

● Draw something in your room or home which you see every day. Only this time draw it from an unusual angle or from a position which makes it look really strange. Draw it very carefully.

● Then measure the object in each direction – up, along and sideways. Write the measurements beside the drawing and ask various people in your family or among your friends to guess what the object is! Can they guess?

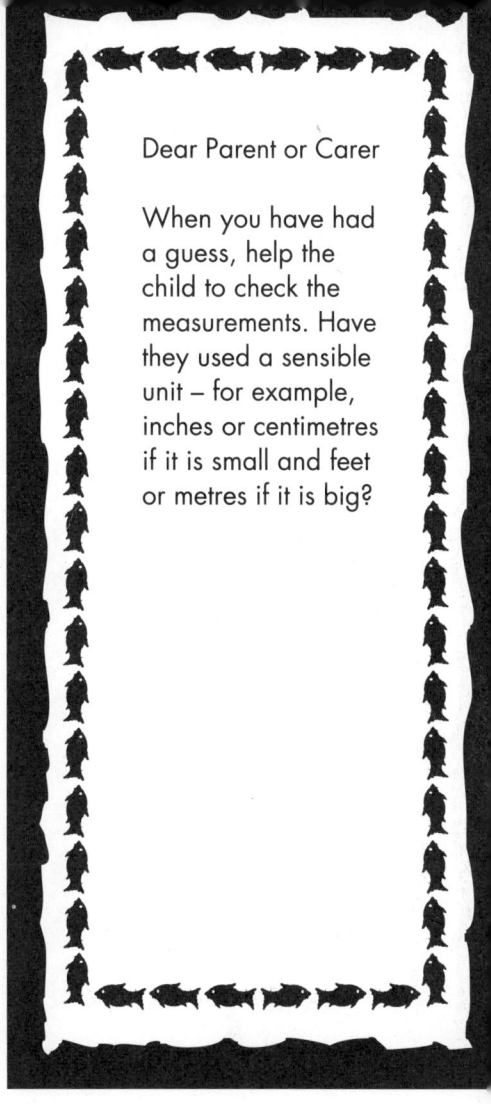

Dear Parent or Carer

When you have had a guess, help the child to check the measurements. Have they used a sensible unit – for example, inches or centimetres if it is small and feet or metres if it is big?

_____ and
child

_____
helper(s)

_____

did this activity together

Holiday activities

Dear Parent or Carer

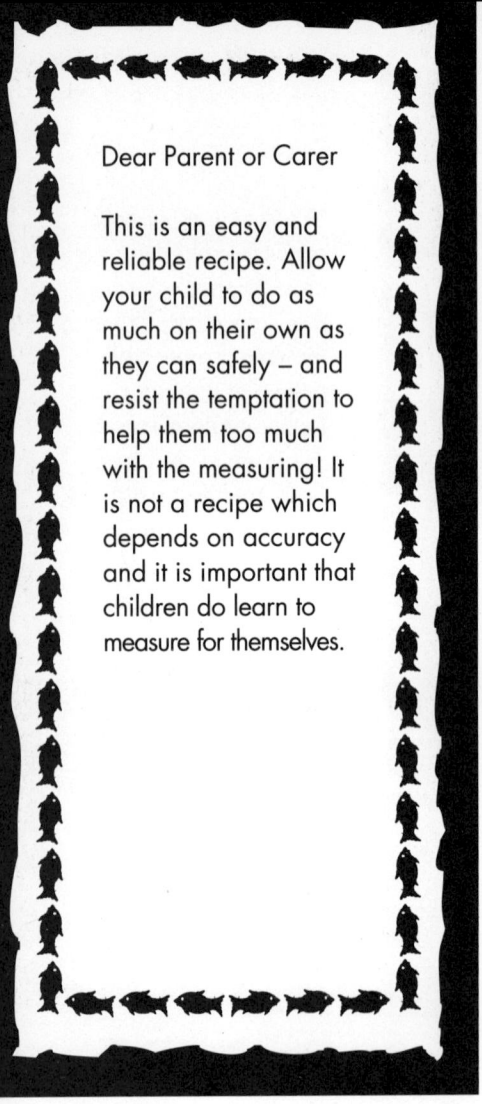

This is an easy and reliable recipe. Allow your child to do as much on their own as they can safely – and resist the temptation to help them too much with the measuring! It is not a recipe which depends on accuracy and it is important that children do learn to measure for themselves.

_____and
child

_____
helper(s)

_____
did this activity together

Holiday activities

Cook away!

You are going to make some delicious biscuits and perform an experiment at the same time! The experiment is to find out if the biscuits will be heavier or lighter than all the ingredients added together.

● Which do you think they will be?

YOU WILL NEED: half a cup of brown sugar, half a cup of white sugar, half a cup of margarine, half a cup of peanut butter, 1 egg, a teaspoon of vanilla essence, one and a quarter cups of self-raising flour, half a teaspoon of bicarbonate of soda, a pinch of salt.

Beat the sugar and margarine together until they are creamy, then add the peanut butter and beat that in. Then add the egg and a teaspoon of vanilla essence and beat again. Finally stir in the self-raising flour, half a teaspoon of bicarbonate of soda and a pinch of salt. Roll the dough into small balls and flatten these on to a greased baking tray. Bake at 375–400°F/190–200°C/Gas 5 for about 15 minutes.

● To perform the experiment, weigh all the ingredients before you use them. (This is easy if you weigh the cup first!)

● Work out the total weight of the ingredients.

● When the biscuits are cooked – make sure they don't burn! – weigh them. Was your guess correct?

impact MATHS HOMEWORK

Racing along

YOU WILL NEED: a toy car and someone to help you measure.

- How far can your car travel in metres? To find out, find a straight smooth place where you can test it.

- Measure one, two and even three metres (if you can). Place markers every ten centimetres so you know where the car gets to.

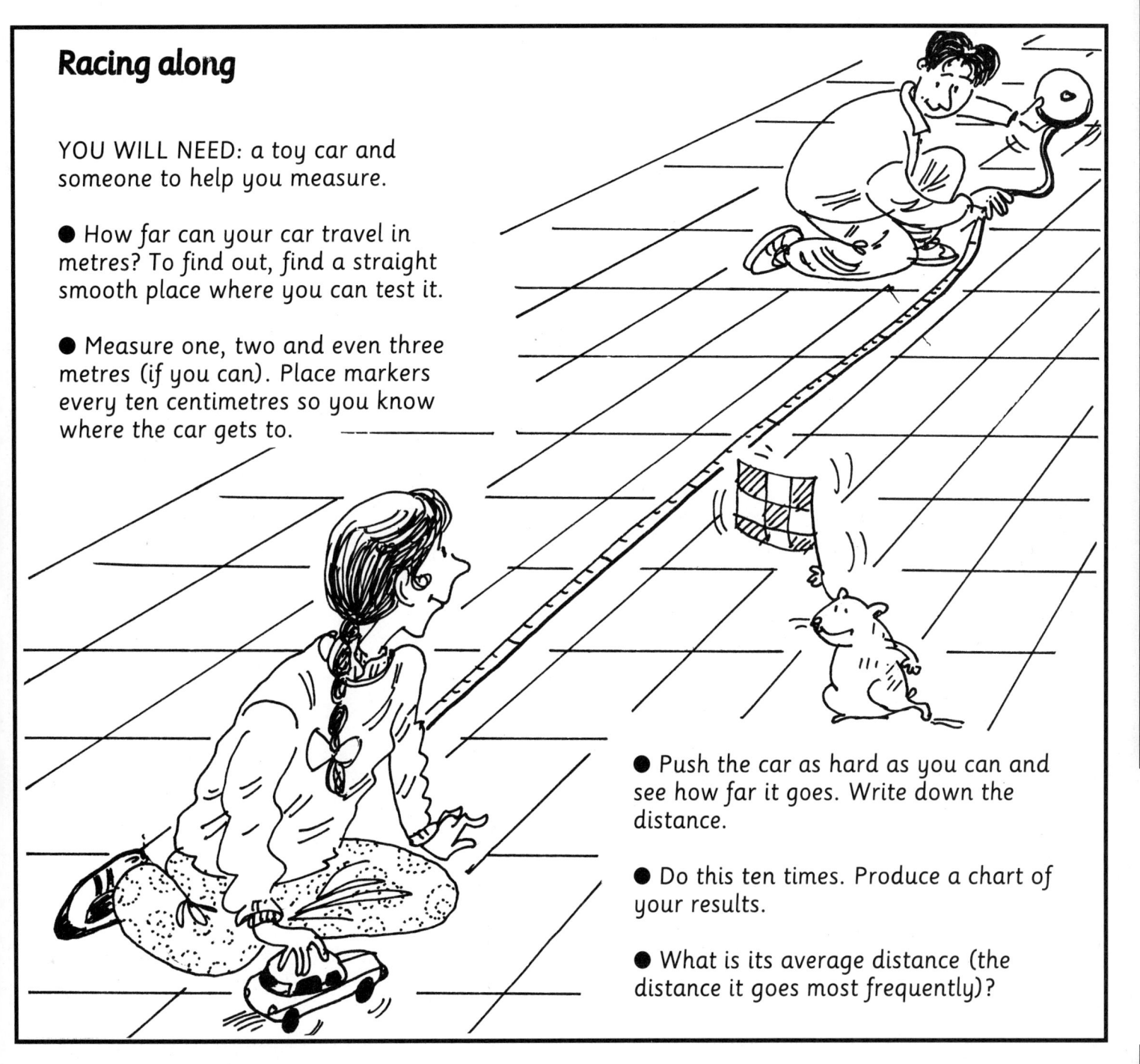

- Push the car as hard as you can and see how far it goes. Write down the distance.

- Do this ten times. Produce a chart of your results.

- What is its average distance (the distance it goes most frequently)?

Dear Parent or Carer

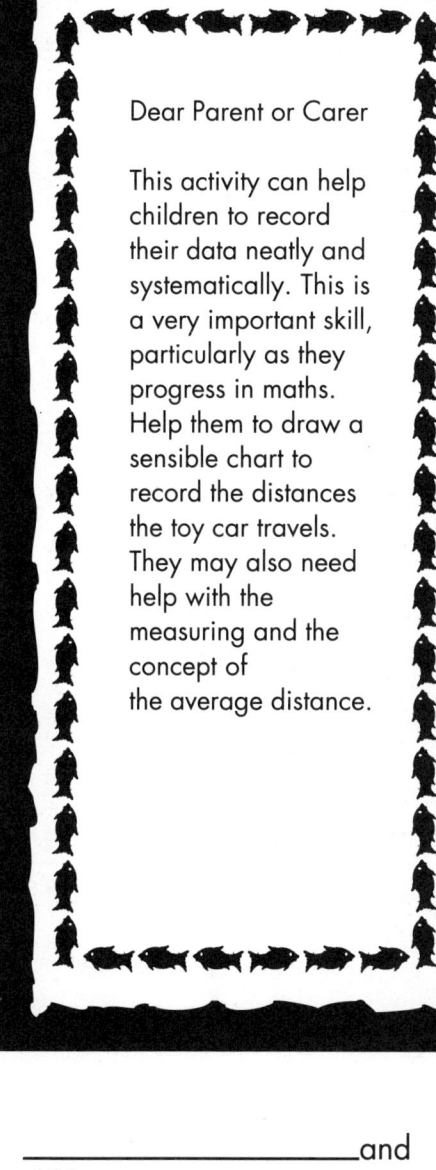

This activity can help children to record their data neatly and systematically. This is a very important skill, particularly as they progress in maths. Help them to draw a sensible chart to record the distances the toy car travels. They may also need help with the measuring and the concept of the average distance.

_____and
child

_____
helper(s)

_____

did this activity together

Holiday activities

Dear Parent or Carer

There is quite a bit of skill in this game. Remember that a multiple is a number that is in the times table of the dice number – for example, if you throw a 2, a multiple of 2 is a number in the two times table. Remember also that squares can stand on their corners and still be squares!

_____ and
child

_____
helper(s)

_____
did this activity together

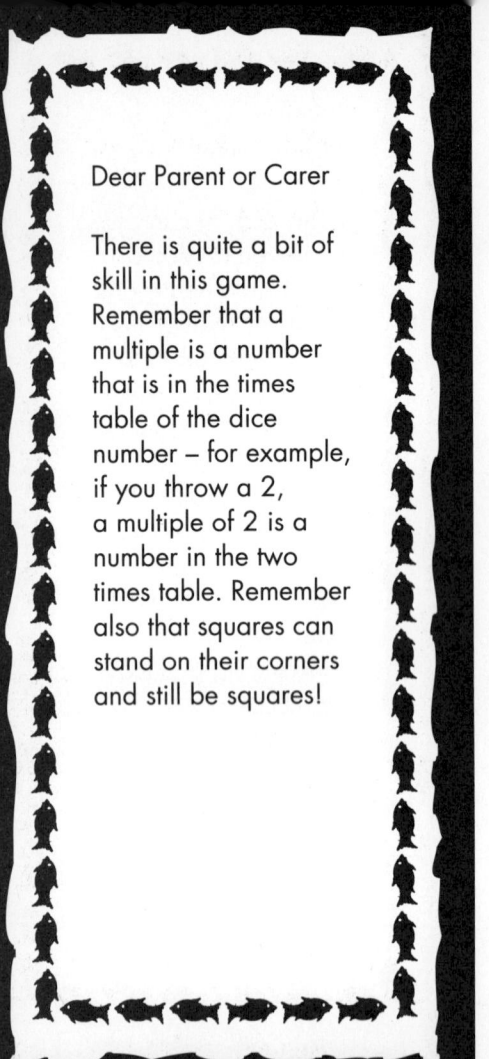

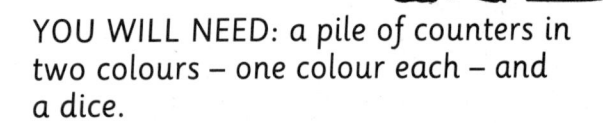

0	1	2	3	4	5	6	7	8	9
10	11	12	13	14	15	16	17	18	19
20	21	22	23	24	25	26	27	28	29
30	31	32	33	34	35	36	37	38	39
40	41	42	43	44	45	46	47	48	49
50	51	52	53	54	55	56	57	58	59
60	61	62	63	64	65	66	67	68	69
70	71	72	73	74	75	76	77	78	79
80	81	82	83	84	85	86	87	88	89
90	91	92	93	94	95	96	97	98	99

Dicey squares

YOU WILL NEED: a pile of counters in two colours – one colour each – and a dice.

● Take it in turns to play. Throw the dice and place a counter of your colour on a space that has a multiple of the number thrown. The first person to cover four spaces that can be joined to make a square is the winner.

Holiday activities

impact MATHS HOMEWORK

Word lengths

- What is the longest word you can think of? How many letters has it?

- If each vowel costs 20p and each consonant costs 15p, how much would it cost to buy this word?

- Can you work out how much it would cost to buy your name?

- How many words can you write that will cost £1?

- Can you write a message to your friend in words so that the whole message costs £1?

Dear Parent or Carer

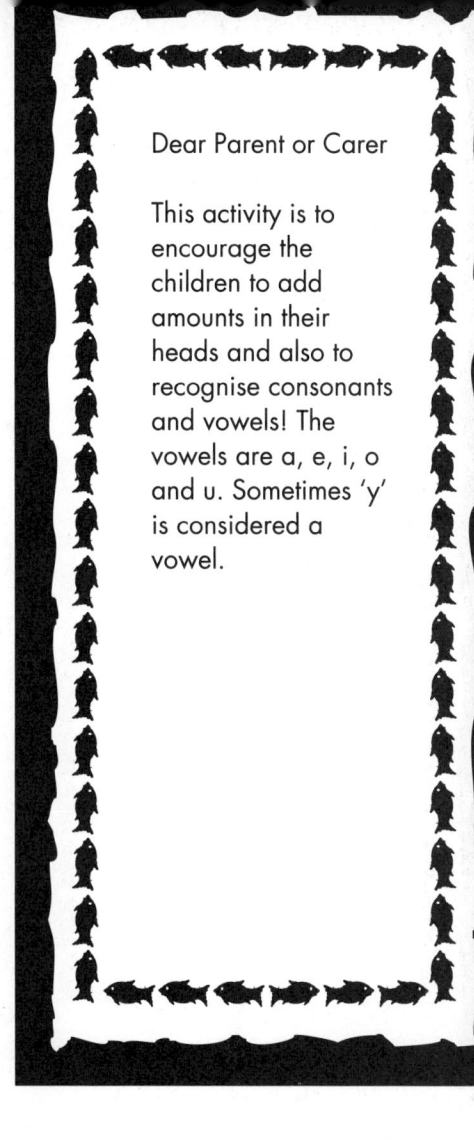

This activity is to encourage the children to add amounts in their heads and also to recognise consonants and vowels! The vowels are a, e, i, o and u. Sometimes 'y' is considered a vowel.

_____and
child

_____
helper(s)

_____

did this activity together

Holiday activities

Dear Parent or Carer

This activity is designed to emphasise the thousands, hundreds, tens and units positions of each number. It is vital that children understand this in order to read really big numbers. If your child finds this activity easy, play with five-figure numbers.

_____ and

child

_____

helper(s)

_____

did this activity together

Holiday activities

Guess the number

YOU WILL NEED: paper, pencils and someone to play with.

● Each of you writes a four-figure number at the top of a piece of paper. Do not show it to your opponent!

● You are trying to guess each other's number.

● You each take it in turns to say a four-figure number out loud. If it has any numbers in it which are in the correct place, then the person must indicate this by saying, 'thousands', 'hundreds', 'tens' or 'units', depending on which one is correct. For example, if I say, '4,568' and the number I am trying to guess happens to be 3,264, my opponent must say 'tens', because the number in the tens column is correct. They do not mention any other column. The first person to guess their opponent's number wins.

● Play again. Who wins this time?

Years and years

- How many years have been lived altogether by the people in your home? Ask everyone to guess first.

- Now work it out. Who was closest?

- How many days is this?

Dear Parent or Carer

It may be helpful to use a calculator for some of the very hard multiplication here, but encourage your child to read the big numbers and say their names out loud.

_____ and
child

_____
helper(s)

_____

did this activity together

Holiday activities

Dear Parent or Carer

This activity is very good for helping children to estimate and get a feel for metric units. It will help if there is a very small prize for the closest estimate!

_____ and
child

_____
helper(s)

_____

did this activity together

Holiday activities

Metres and metres

How many metres is it to your local shops? You can pace it out and find out. A fairly large pace is half a metre.

● Find a tape measure or ruler and cut a strip of newspaper half a metre long. Now practise walking with this size paces!

● Guess how many metres it is to the shops. Ask as many people as possible in your home to guess as well. Perhaps there might be a prize for the person whose guess is closest.

● Pace it out carefully. Remember to divide the number of paces by 2 to find the number of metres.

● Whose guess was closest?

impact MATHS HOMEWORK

Washing up!

- Who does the washing up in your house? Is it always the same person? Is it a machine?

- Choose a large meal and ask everyone to estimate the number of minutes it will take to wash up – or load the dishwasher!

- Write down all their guesses.

- When the washing up starts, time it accurately.

- How many minutes did it take?

- Who was closest? Perhaps they get a holiday from washing up and the person who said the worst estimate washes up next!

Dear Parent or Carer

Estimating how long things take is quite a difficult skill. Actually, some adults aren't too skilled at this either! Help your child to time accurately – either by counting the number of minutes the big hand has moved or by counting the minutes passing on a digital clock.

_____and
child

_____
helper(s)

_____

did this activity together

Holiday activities

Dear Parent or Carer

Playing games is one of the best ways to practise mental arithmetic and tables facts. Make the game as much fun as possible and play it often!

_____ and
child

_____
helper(s)

_____

did this activity together

Holiday activities

Loads of money!

YOU WILL NEED: a counter each, a dice, the board provided, a pencil and paper each to keep track of your money, a pack of cards with the face (picture) cards removed, and someone to play with.

- Place the cards face down in the space labelled on the board.

- Each time you throw the dice, you take a card and move along the spaces on the track according to the number on the dice.

- Using the number on the card, follow the instructions on the space you land on. If the space you land on has a £ sign, you score the amount in pounds. If it has a 50p sign, you score the amount in 50p coins. If it has a 10p sign, you score the amount in 10p coins. All arithmetic must be agreed by all the players!

- Using your pencil and piece of paper, keep a record of each score you make. The person who has collected the most money at the end is the winner.

impact MATHS HOMEWORK

Loads of Money

Place the cards face down here

Board spaces (from START):

- START
- Multiply card number by 3. £
- Multiply card number by 6. 50p
- Add 10 to card number. £
- Multiply card number by 4. 50p
- Multiply card number by 10. 10p
- Add 13 to card number. £
- Double card number. 50p
- Multiply card number by 9. 10p
- Multiply card number by 9. £
- Add 1 to card number and double. 50p
- Multiply card number by 8. 10p
- Double card number. £
- Add 27 to card number. £
- Add 3 to card and x by 7. 50p
- Add 2 and x 8. 10p
- Add 5 and double. 50p
- Multiply by 7. £
- Add 19 and double. 50p
- Multiply by 5 and add 10. 10p
- Multiply by 8. 50p
- Add 13 and double. £
- Multiply by 6 and halve. £
- Add 2 and multiply by 10. £
- Add 10 and halve. 50p
- Double and add 12. £
- Multiply by 6. 50p
- Multiply by 4. £
- Multiply by 10. 10p

Holiday activities

Dear Parent or Carer

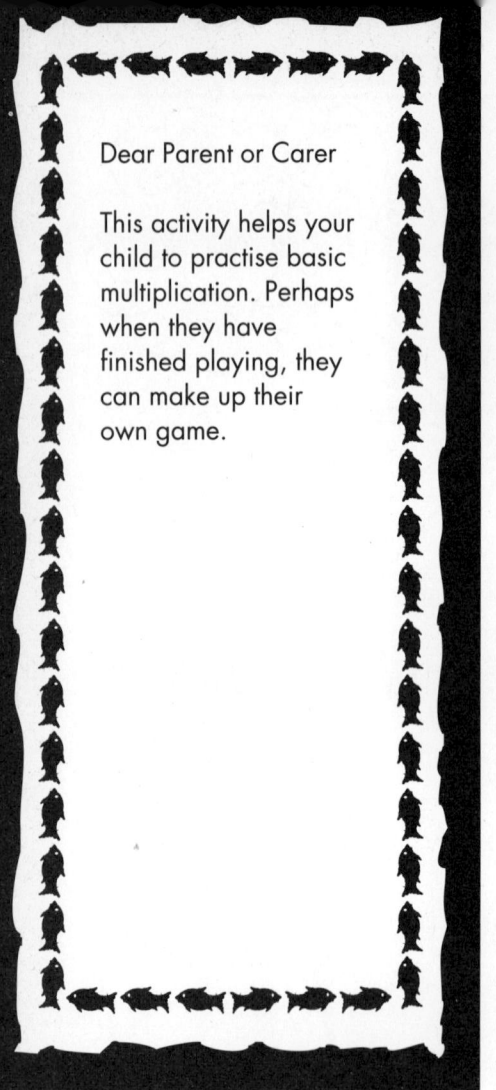

This activity helps your child to practise basic multiplication. Perhaps when they have finished playing, they can make up their own game.

_____ and
child

_____
helper(s)

_____
did this activity together

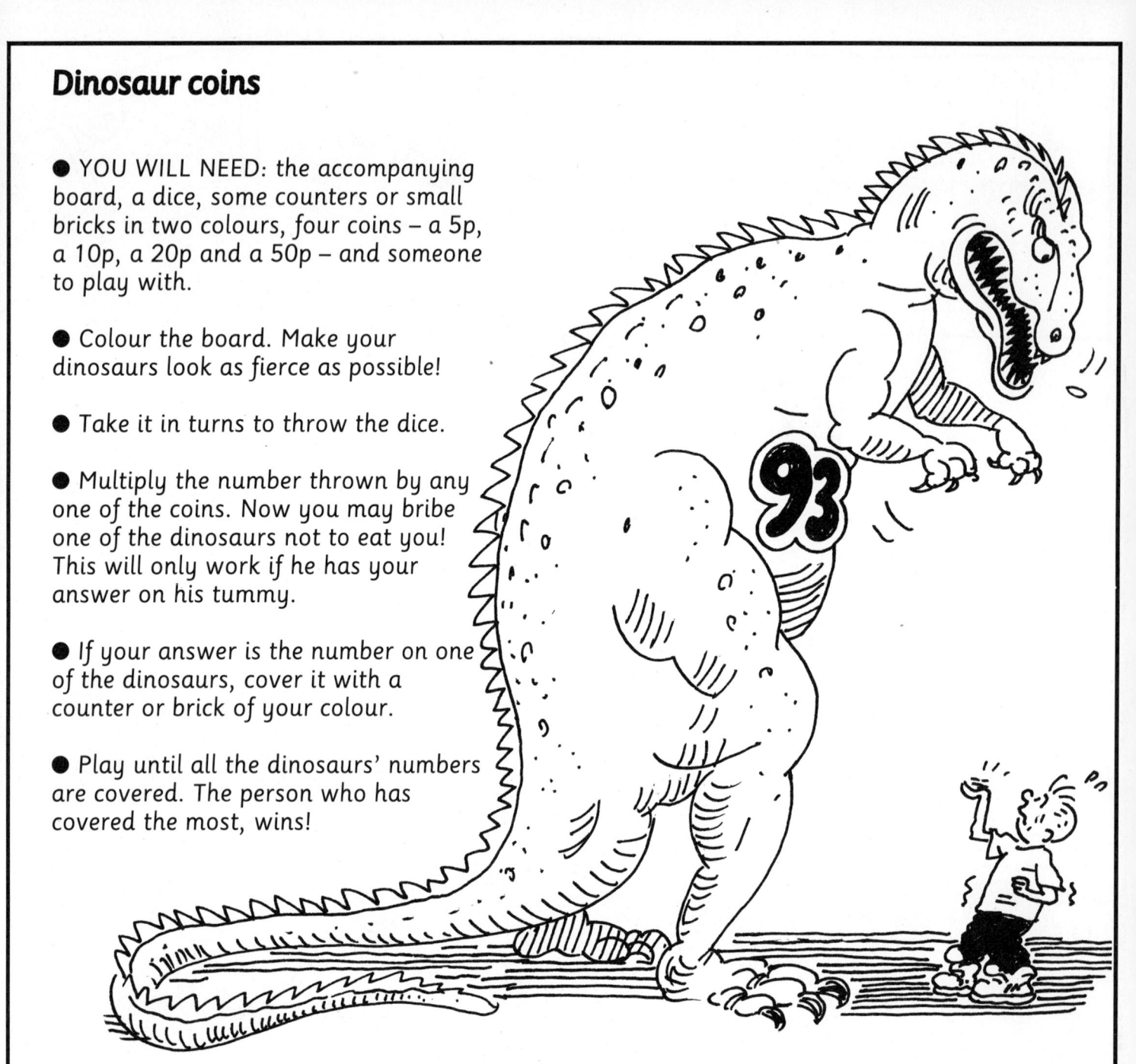

94 **Holiday activities**

Dinosaur coins

- YOU WILL NEED: the accompanying board, a dice, some counters or small bricks in two colours, four coins – a 5p, a 10p, a 20p and a 50p – and someone to play with.

- Colour the board. Make your dinosaurs look as fierce as possible!

- Take it in turns to throw the dice.

- Multiply the number thrown by any one of the coins. Now you may bribe one of the dinosaurs not to eat you! This will only work if he has your answer on his tummy.

- If your answer is the number on one of the dinosaurs, cover it with a counter or brick of your colour.

- Play until all the dinosaurs' numbers are covered. The person who has covered the most, wins!

impact MATHS HOMEWORK

Dinosaur coins

Dear Parent or Carer

This activity helps children to practise addition. Be sure to try out the one they make up!

_____and
child

_____
helper(s)

_____
did this activity together

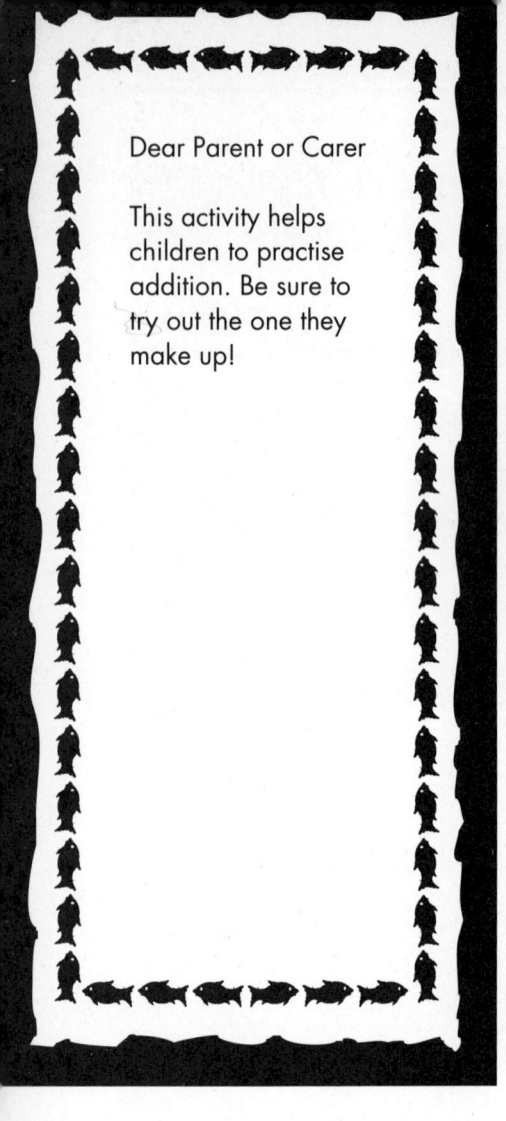

96   **Holiday activities**

Space hopping

7	5	13	7	29	15
21	14	16	9	23	31
32	20	8	45	11	19
18	13	10	24	22	17
32	26	30	12	6	27
28	1	25	5	10	4

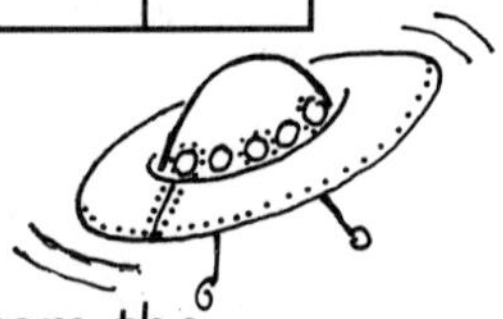

● Can you find a route from the monster to his spaceship which adds up to exactly 100?

● Add the numbers on each square you touch. You may move horizontally, diagonally or vertically.

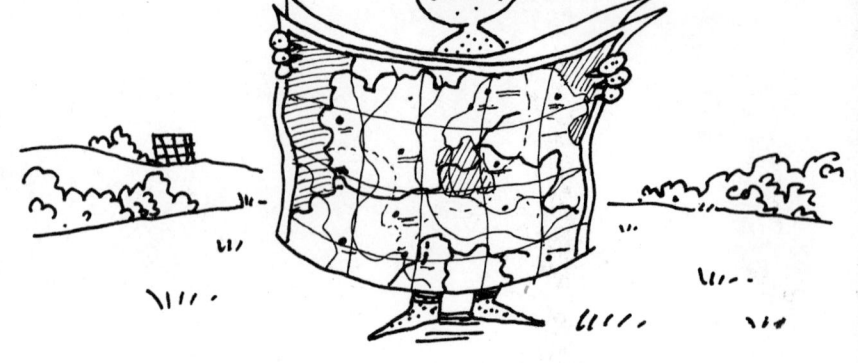

● Now, can you invent a new number square and a new route adding up to 100 for someone else to try?

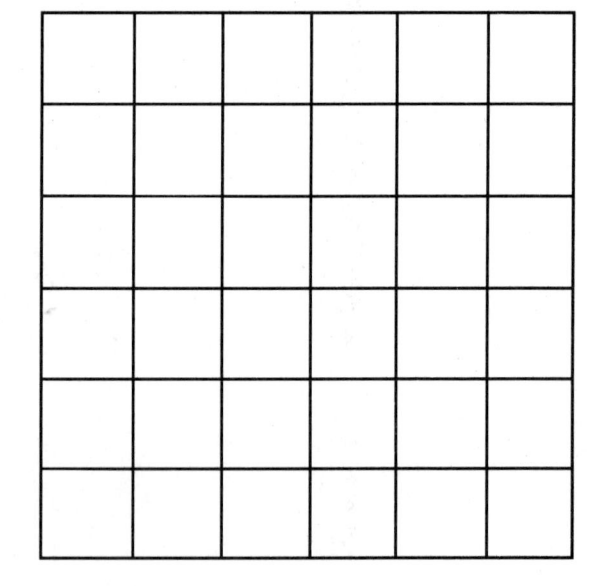

impact MATHS HOMEWORK